Ionization and Ion Transport (Second Edition)

A primer for the study of gas discharges and plasmas

Online at: https://doi.org/10.1088/978-0-7503-3991-9

IOP Series in Plasma Physics

Series Editor

Richard Dendy

Culham Centre for Fusion Energy and the University of Warwick, UK

About the series

The IOP Plasma Physics ebook series aims at comprehensive coverage of the physics and applications of natural and laboratory plasmas, across all temperature regimes. Books in the series range from graduate and upper-level undergraduate textbooks, to research monographs and reviews.

The conceptual areas of plasma physics addressed in the series include:

- Equilibrium, stability, and control.
- Waves: fundamental properties, emission, and absorption.
- Nonlinear phenomena and turbulence.
- Transport theory and phenomenology.
- Laser–plasma interactions.
- Non-thermal and suprathermal particle populations.
- Beams and non-neutral plasmas.
- High energy density physics.
- Plasma–solid interactions, dusty, complex, and non-ideal plasmas.
- Diagnostic measurements and techniques for data analysis.

The fields of application include:

- Nuclear fusion through magnetic and inertial confinement.
- Solar–terrestrial and astrophysical plasma environments and phenomena.
- Advanced radiation sources.
- Materials processing and functionalisation.
- Propulsion, combustion, and bulk materials management.
- Interaction of plasma with living matter and liquids.
- Biological, medical, and environmental systems.
- Low-temperature plasmas, glow discharges, and vacuum arcs.
- Plasma chemistry and reaction mechanisms.
- Plasma production by novel means.

A full list of titles published in this series can be found here: https://iopscience.iop.org/bookListInfo/iop-plasma-physics-series.

Ionization and Ion Transport (Second Edition)

A primer for the study of gas discharges and plasmas

David B Go
Department of Aerospace and Mechanical Engineering, University of Notre Dame, Notre Dame, IN, USA

IOP Publishing, Bristol, UK

ISBN 978-0-7503-3991-9 (ebook)
ISBN 978-0-7503-3989-6 (print)
ISBN 978-0-7503-3992-6 (myPrint)
ISBN 978-0-7503-3990-2 (mobi)

DOI 10.1088/978-0-7503-3991-9

Version: 20221201

IOP ebooks

British Library Cataloguing-in-Publication Data: A catalogue record for this book is available from the British Library.

Published by IOP Publishing, wholly owned by The Institute of Physics, London

IOP Publishing, No.2 The Distillery, Glassfields, Avon Street, Bristol, BS2 0GR, UK

US Office: IOP Publishing, Inc., 190 North Independence Mall West, Suite 601, Philadelphia, PA 19106, USA

Dedicated to my Beth, Indiana, and Donovan.
In loving memory of Elizabeth D Go.

Contents

Preface **xi**

Acknowledgement **xiv**

Author biography **xv**

1 Introduction **1-1**

1.1 Overview 1-1
1.2 Classification of gas discharges 1-3
1.3 Summary 1-6

Part I Ionization

2 Foundations from gas dynamics **2-1**

2.1 The atom, the molecule, and excited states 2-2
2.2 The statistics of a gas 2-6
2.3 Kinetics 2-9
2.3.1 Collision theory 2-9
2.3.2 Equilibrium behavior 2-12
2.4 Summary 2-15
References 2-16

3 Elementary electron behavior **3-1**

3.1 Electron-driven gaseous reactions 3-1
3.1.1 Elastic collisions 3-2
3.1.2 Excitation collisions 3-2
3.1.3 Ionization collisions 3-3
3.2 Electron–neutral collisions and the collision cross-section 3-4
3.2.1 The nature of electron collisions 3-4
3.2.2 Differential collision cross-section 3-4
3.2.3 Elementary collision dynamics 3-6
3.2.4 Realistic collision cross-sections 3-9
3.3 Electron energy distribution 3-10
3.4 Electron collision frequency and reaction rates 3-13
3.5 Summary 3-15
References 3-16

4 Gaseous ionization processes **4-1**

4.1 Electron-impact ionization 4-1

4.1.1 Temporal electron avalanche 4-2

4.1.2 Spatial electron avalanche 4-2

4.1.3 Townsend's first ionization coefficient α 4-5

4.2 Ion-impact ionization 4-6

4.3 Photoionization 4-7

4.4 Thermal ionization 4-8

4.5 Step ionization 4-8

4.6 Penning ionization 4-9

4.7 Electron attachment 4-10

4.8 Summary 4-10

References 4-11

5 Electrode processes **5-1**

5.1 Importance of electron emission 5-1

5.2 Foundational concepts 5-2

5.2.1 The solid state 5-2

5.2.2 Fermi–Dirac distribution and the work function 5-4

5.3 General electron emission 5-5

5.3.1 Thermionic emission 5-7

5.3.2 Photoemission 5-8

5.3.3 Field emission 5-9

5.4 Secondary emission 5-11

5.4.1 Ion-induced secondary emission 5-11

5.4.2 Secondary emission coefficient γ 5-15

5.4.3 Metastable-induced secondary emission 5-16

5.5 Summary 5-17

References 5-17

Part II Ion transport

6 Foundations from general transport theory **6-1**

6.1 Basics of transport 6-1

6.1.1 Application of kinetics to transport 6-3

6.2 Transport equations 6-5

6.2.1 Continuum perspective—scalar transport equation 6-5

	6.2.2 Kinetic perspective—Boltzmann transport equation	6-8
	6.2.3 From the Boltzmann transport equation to macroscopic properties	6-10
6.3	Summary	6-11
	References	6-12
7	**Transport equations for gas discharges**	**7-1**
7.1	Equations for the charged particles	7-1
	7.1.1 Particle/charge conservation	7-1
	7.1.2 Momentum conservation	7-3
	7.1.3 Energy conservation	7-4
	7.1.4 Electric field	7-5
7.2	The drift–diffusion approximation	7-6
7.3	Summary	7-8
	References	7-9
Part III	**Gas discharges and plasmas**	
8	**Discharge initiation and plasma formation**	**8-1**
8.1	Townsend behavior below the breakdown threshold	8-1
8.2	Pressure times distance and reduced electric field scaling	8-5
8.3	Breakdown	8-6
	8.3.1 Townsend breakdown	8-7
	8.3.2 Streamer breakdown	8-10
8.4	Summary	8-13
	References	8-14
9	**The basic discharge structure**	**9-1**
9.1	The sheath and the plasma	9-1
9.2	The physics of the sheath	9-3
	9.2.1 Debye length	9-4
	9.2.2 The potential field in the sheath	9-5
9.3	Summary	9-7
	References	9-8
10	**The classical dc discharge**	**10-1**
10.1	The basics of electrical circuits	10-1

10.2 dc discharge circuit and operation 10-3
10.2.1 The canonical current–voltage curve 10-5
10.2.2 The discharge circuit 10-6
10.3 Properties of the dc glow discharge 10-8
10.3.1 The cathode fall 10-10
10.4 Summary 10-12
References 10-13

11 Whence and whither 11-1

11.1 What was learned 11-1
11.2 What is next 11-2

Preface

The world of gas discharges and plasmas, often called plasma science and engineering, is one that is exciting, challenging, extremely broad, and full of incredible promise for societal impact. If you were to ask anyone active in the field, they would not hesitate to describe the interesting science and the wide range of technologies that do, will, and could depend on gas discharges and plasmas. If they work at the high temperature end of the plasma spectrum, they might describe to you the potential of fusion to solve all our energy problems or how plasmas are important in understanding our Sun, our Solar System, our Galaxy, and really the entire Universe (99% of the known matter in the entire world is essentially a plasma). If they work with low-temperature, non-equilibrium plasmas, they would probably become very animated as they describe using a plasma to heal a wound or treat cancer ('plasma medicine'), using a plasma to fertilize plants and increase crop growth ('plasma agriculture'), or developing reactors to convert our most abundant pollutant, carbon dioxide, into useful chemicals ('plasma catalysis')[1]. The truth is, those of us that work with plasmas do it because, while yes they are impactful, they are just incredibly cool[2].

But the challenge is: how does one 'get into' plasma science and engineering? It is not a field that is well known to most students in the way that being an engineer, chemist, or physicist is. Most colleges and universities worldwide do not have a Department of Plasma Science and Engineering or offer degrees in gas discharges or plasma science and engineering. You would likely be hard pressed to find a course on plasma, in particular at the undergraduate level, in most science and engineering schools. In fact, most *people* likely do not know that plasma science and engineering even exists as a field. If you were to take a random sampling of the general public, most identify *plasma* with *blood plasma*. And because it is a relatively niche and therefore not easily accessible field, even for those of us who work with, conduct research on, or otherwise study gas discharges and plasmas, it is not necessarily easy to learn the field, whether that be at the basic level of understanding the terminology to the advanced level of knowing the mathematics that describe plasma behavior. Further, because many of us encounter plasma science and engineering from vastly different backgrounds, ranging from mechanical and electrical engineering to chemistry and physics, our entry points into the field may be at very different levels of background knowledge.

The reason that I wrote this primer, and now am offering this second edition, is to create an entry point for students and scholars who are just beginning their journey into the world of plasma science and engineering. It is intentionally short and (hopefully) readable, but without sacrificing technical information. I also have tried not to assume the background of the 'average' reader other than they are likely an

[1] If you haven't noticed, the plasma community isn't great at naming things. They just like adding 'plasma' in front.

[2] And in principle, those of us that work with gas discharges and plasmas are also cool by the transitive property.

upperclass student pursuing an undergraduate degree or a first- or second-year graduate student in some STEM discipline[3]. (Or perhaps they are a staff engineer or hobbyist with some technology background.) The purpose of this text is to introduce engineering and science students to the basic underlying physics and chemistry concepts that form the foundation of plasma science and engineering. As topics can range from statistical thermodynamics and gas dynamics to fluid dynamics and solid-state physics, it is unlikely that most readers have backgrounds in all these areas. Therefore, I have tried to both provide general information in these areas as well as specific content that applies directly to gas discharges and plasmas. In my own teaching, this text is based on my notes for an entry-level graduate course I have taught on gas discharges and plasmas, covering roughly the first two-thirds of a semester.

In order to understand the structure of the text (and the title), it is important to understand what a gas discharge or plasma is—very generally speaking, a plasma is a gas that consists of free electrons and ionized gas particles (i.e. missing or having an extra electron). Therefore, fundamentally, understanding the field depends on understanding how ions are created—‘ionization’—and how they move—‘ion transport’; hence the name of this book. While I retain the same structure of the first edition, organizing the topics into a part I on ionization and part II on ion transport (with minor edits and corrections), new to this second edition is a part III that covers more specific details of gas discharges, including how they are formed (breakdown), some of their basic qualities, and the simple, yet ubiquitous, glow discharge. Again, as in the first edition, the focus is on topics mostly relevant to non-equilibrium, low-temperature gas discharges and plasmas rather than to the extremely broad field of plasma science more generally.

While part III certainly introduces more detailed gas discharge topics, as a primer this text is again intentionally short and thus incomplete. As such this book is not meant to comprehensively cover plasma physics, including the many different modes of plasma operation or the wide range of plasma technologies. The reader is instead directed to the many other useful plasma resources that are available, including an incredible set of journal articles curated by the Institute of Physics journal *Plasma Sources Science and Technology* called ‘Foundations of low-temperature plasmas and their applications’, edited by L L Alves and A von Keudell[4]. This set of articles covers a wide variety of plasma topics, both fundamental and applied, with the express intent of being educational, rather than a review or presentation of new research. They would make a wonderful complement to this book, where they present many topics in more depth while this text (hopefully) fills in some of the missing background that they leave out.

As with the first edition, I was influenced by a number of more comprehensive and detailed texts as I was writing this second edition. I have tried my utmost to reference them appropriately throughout, and I owe a debt of gratitude to them all.

[3] STEM is an acronym for Science, Technology, Engineering, and Mathematics.

[4] Available at: https://iopscience.iop.org/journal/0963-0252/page/Foundations-of-low-temperature-plasmas-and-their-applications

I encourage all readers to move on to these texts once they complete this primer for a much deeper understanding of non-equilibrium, low-temperature plasmas.

- Nasser E 1971 *Fundamentals of Gaseous Ionization and Plasma Electronics* (New York: Wiley-Interscience)
- Raizer T P 1991 *Gas Discharge Physics* (Berlin: Springer)
- Lieberman M A and Lichtenberg A J 2005 *Principles of Plasma Discharges and Materials Processing* 2nd edn (New York: Wiley-Interscience)
- Fridman A 2008 *Plasma Chemistry* (New York: Cambridge University Press)
- Fridman A and Kennedy L A 2011 *Plasma Physics and Engineering* 2nd edn (Boca Raton, FL: CRC Press)

I stumbled into the world of plasma science and engineering almost by accident in graduate school when I was offered the opportunity to work on a research project that involved electrohydrodynamic flow. It was not a field I ever planned or imagined myself being in, but now—more than 15 years later—I cannot imagine myself *not* working with gas discharges and plasmas. It is my sincere hope that the readers of this text similarly have the chance to pursue a very fulfilling career in plasma science and engineering. I am pleased that this book may play a small part in that journey.

David B Go
University of Notre Dame
Notre Dame, Indiana 46556
United States of America

Acknowledgement

In the first edition of this book I thanked my friends and colleagues in the plasma research community as well as the University of Notre Dame students that braved taking my graduate-level course Ionization and Ion Transport, which as you might surmise, formed the namesake of this text. I continue to be very grateful for all my thoughtful and fruitful discussions at meetings, conferences, and in the classroom that have entrenched my fondness for plasma science.

In this second edition, though, I want to specifically call out the graduate students and my collaborators at the University of Notre Dame that have worked directly with me on plasma research and played such an incredible part in my career as a researcher and educator. The majority of them have read some version of this book, whether it be my earliest class notes, the first edition, or drafts of this second edition, and for that I am appreciative. Some have even contributed data that went into figures in this second edition. They have also enriched my world in numerous ways—both in the lab and out—that have made my career in academia so fulfilling. I am eternally grateful that I have been able to share the joy of discovery, the joy of learning, and the joy of plasma with all of them.

Author biography

David B Go

David B Go is the Viola D Hank Professor of Aerospace and Mechanical Engineering at the University of Notre Dame in Notre Dame, Indiana, USA. After growing up close to the university, he graduated with a BS in Mechanical Engineering from the University of Notre Dame. He then spent several years in the Edison Engineering Development Program at General Electric Aviation (formerly Aircraft Engines), concurrently earning an MS in Aerospace Engineering from the University of Cincinnati. Professor Go ultimately received his PhD in Mechanical Engineering from Purdue University and joined the faculty of the University of Notre Dame in 2008. He has published widely in the areas of plasma science and gas discharges, heat transfer and fluid dynamics, and chemical analysis, authoring numerous journal articles, conference papers and presentations, multiple book chapters, and obtaining several patents. Professor Go has been recognized with the Air Force Office of Scientific Research Young Investigator Research Award, the National Science Foundation CAREER award, the IEEE Nuclear and Plasma Sciences Society Early Achievement Award, the inaugural Electrostatics Society of America Rising Star Award, and as an inaugural winner of the Electrochemistry Society Toyota Young Investigator Fellowship. He is a Fellow of the American Society of Mechanical Engineers (ASME) and a Senior Member of the Institute of Electrical and Electronics Engineers (IEEE).

IOP Publishing

Ionization and Ion Transport (Second Edition)
A primer for the study of gas discharges and plasmas
David B Go

Chapter 1

Introduction

1.1 Overview

The main focus of this book is to gain an understanding of gaseous ionization and ion transport in relation to gas discharges and plasmas. However, for the uninitiated, even these terms may be unfamiliar. Therefore, it is best to begin with a short discussion of the basic concepts and basic terminology.

What is a gas discharge? A gas discharge is essentially an electrical flow called a current across an interstitial gap of gas—turning the insulating gas into a conductor. (For this reason, it was originally associated with the discharge of a capacitor.) In a laboratory or artificial setting, this process occurs by inputting energy, typically through the application of an electric or magnetic field, to generate freely moving charged particles (consisting of electrons and ions) in the gas, making it conductive. However, discharges occur in nature as well—with lightning and static electricity being two very common examples. For the purposes of this book, we will focus on laboratory-generated discharges. There are many different properties of a gas discharge, and we classify them based on these properties. Plasma and gas discharges are closely related and both consist of gaseous ions, although it is best to call gas discharges a subset of plasmas, in general.

What is an ion? A gas discharge requires the generation of gas-phase electrons and ions that are free to move, which is called *ionization*. But what is an 'ion'? An ion is an atom or molecule where the total number of electrons is not equal to the number of protons. Therefore, the atom or molecule carries either excess positive charge (number of protons > number of electrons) or excess negative charge (number of electrons > number of protons)—thus it is either a positive or negative ion. Electrons themselves are often called negative ions, but for the purposes of this manuscript we will delineate negative ions as an atom or molecule with net negative charge and electrons simply as electrons. (Similarly, we will call positive ions an atom or molecule with net positive charge and protons just protons.)

doi:10.1088/978-0-7503-3991-9ch1

Why do these ions move? Consider basic electronics. When a voltage is applied across a conductor such as a metal, a flow of electrons is generated. This is, in a nutshell, electricity. If we break it down, we see that electricity—or the flow of electrons—requires the medium to be conductive, i.e. having charges that are free to move throughout the medium, from atom to atom, and it also requires some force to induce those charges to move. Let us look a little bit closer, but still at an elementary level.

The application of voltage induces the electrons to move, so naturally the questions is: what is voltage? Voltage is merely *electric potential difference*. In this example, the electrons want to go from 'high' potential energy to 'low' potential energy—in the same way that a ball wants to roll down a hill due to gravitational potential energy. The electrostatic potential energy is the voltage or electric potential (Φ) multiplied by the charge (q)[1] of the matter in question,

$$U = q\Phi. \tag{1.1}$$

Now, all potential energy manifests itself as a force on matter, and for conservative forces (such as electrostatic forces), this is determined by the negative of the gradient of the potential energy, or $\vec{F} = -\vec{\nabla}U$. For simplicity, let us only consider one dimension; applying this relationship, the force on charge q is therefore

$$F_x = -\frac{dU}{dx} = -q\frac{d\Phi}{dx} = qE_x. \tag{1.2}$$

Here, E_x is the electric field and F_x is the Coulombic force[2], both in the $\hat{x}$-direction. From this, it is easy to see that the electric field (force) is related to the electric potential (energy) by a derivative. More generally, this relationship is a gradient,

$$\vec{E} = -\vec{\nabla}\Phi. \tag{1.3}$$

The potential energy *distribution* is governed by Poisson's equation[3] or

$$\nabla^2\Phi = -\frac{(n_+ - n_-)}{\epsilon_0}, \tag{1.4}$$

where ϵ_0 is the permittivity of free space[4]. The variables n_+ and n_- are the charge densities (or concentrations) of positive ions and negative ions/electrons,

[1] The unit charge q takes a value of $q = 1.602\,176\,5 \times 10^{-19}$ C. A single proton has a charge of $+q$ and similarly, a single electron has a charge of $-q$. In most cases, we will assume that ions have a single charge, i.e. +1 or −1, but there are certainly cases, particularly at higher pressures, where ions with more charges are possible.

[2] Named after Charles-Augustin de Coulomb, a French physicist and engineer and a pioneer in electrostatics. The unit of electric charge is named the coulomb (and takes the symbol C).

[3] Named after Siméon Poisson, another French pioneer in the field of electrostatics and electromagnetics, as well as a variety of other fields.

[4] $\epsilon_0 = 8.85 \times 10^{12}$ F m^{-1}.

respectively[5]. (That is, $n = qN$, where N is the number density (or concentration) of each species[6].)

So in an electric circuit, the motion of electrons is due to the force from the applied potential energy, i.e. voltage difference. This induces an *electrostatic field* that drives the charges creating the *current, i*, which is nothing more than the time derivative of charge

$$i = \frac{dq}{dt}. \tag{1.5}$$

The unit of current is charge per unit time (C s^{-1}), also called an ampere (A). Typically, we refer to a *current density* with the symbol j to indicate current through an area, with a unit of A m^{-2}.[7]

Ions in the gas phase have charge and therefore also respond to an applied voltage to form what is called a discharge current. Therefore, one can conclude that the two most important facets of understanding gas discharges are:

1. The creation of charges in the gas (*ionization*) to make it conductive.
2. The motion of charges through the gas (*ion transport*) to form a current.

1.2 Classification of gas discharges

How are gas discharges classified? Discharges consist of different types of particles—the background gas (neutral charge), excited gas particles (also neutral charge), ionized gas particles (positive and negative ions), and electrons. Each of these species will have some average kinetic energy. Generally, energy[8] ε can also be described by a temperature[9] T using the Boltzmann[10] constant k_B,

$$\varepsilon = k_B T, \tag{1.6}$$

where $k_B = 1.3807 \times 10^{-23}$ J K^{-1} = 8.617 343 $\times 10^{-5}$ eV K^{-1}. This effective 'temperature' plays an important role in characterizing the energy in the discharge, and thus on how we classify discharges. In fact, gas discharges are generally classified into two categories based on the energy of the major particles in the gas.

Non-equilibrium, low-temperature discharges, also called non-thermal plasmas, are gas discharges where the temperature (energy) of the electrons is much higher than the temperature (energy) of the gas neutrals and ions. That is, $T_{elec} \sim 10^4$ K $\gg T_{gas} \sim T_{ion} \sim 10^2$–$10^3$ K (or in eV, $T_{elec} \sim 1$–10 eV $\gg T_{gas} \sim T_{ion} \sim 0.01$ eV). Most industrial and laboratory gas discharges are non-equilibrium, low-temperature, with

[5] The units of n are typically C m^{-3}.

[6] Number densities have units of number per m^3, or more simply m^{-3}. Often, however, number densities are reported in cm^{-3} rather than m^{-3}, but both are acceptable.

[7] Current density can also be thought of as a current flux—the flow of current through an area—but it is convention to call it current density.

[8] In plasma science and engineering, energies are reported in electronvolts (eV) which is defined as the work done on an electron by accelerating it through a potential difference of 1 V. In the conventional units of joules (J) often used in engineering, 1 eV = 1.602 176 6208 $\times 10^{-19}$ J.

[9] In this text, temperature has units of kelvins (K).

[10] Named after the Austrian physicist and philosopher Ludwig Eduard Boltzmann.

common examples including glow discharges, corona discharges, dielectric barrier discharges (DBDs), and radio frequency (RF) discharges. One notable characteristic of non-equilibrium, low-temperature discharges is that the gas temperature can be as low as room temperature, which implies that the gas is not hot enough to harm human flesh. In fact, the emerging field of plasma medicine shows that with the right configuration, non-equilibrium, low-temperature discharges can be used to actually treat skin ailments without harming the patient.

Equilibrium discharges, also called thermal plasmas, are when the temperature (energy) of the electrons and the temperature (energy) of the gas and ions are roughly equal and well above room temperature; $T_{elec} \sim T_{gas} \sim T_{ion} \sim 10^4$ K or much greater. Arc discharges are a prime example of an equilibrium discharge. Our primary focus in this text will be on ionization and ion transport processes relevant to non-equilibrium, low-temperature gas discharges, but it is important to note that many of the basic fundamentals we will discuss apply just as well to thermal plasmas.

Discharges are often characterized by their *degree of ionization* as well, which is simply the percentage of the gas that is ionized. For non-equilibrium, low-temperature gas discharges, this is very small, often no more than 0.1%, which means that over 99.9% of the gas remains neutral or uncharged. Yet even this small amount of ionization is enough to make the gas conductive. For thermal plasmas, it really depends on the regime of operation, and the degree of ionization can range from several percent to 100%, which is referred to as fully ionized. This large disparity, and the use of the term 'thermal plasma' raises an important question.

What is a plasma as opposed to a gas discharge? The first hurdle in characterizing a plasma or gas discharge is defining exactly what a plasma is, in particular relative to a gas discharge. As we learned above, a gas discharge is the conversion of an insulating gas into a conductive gas, most often by the application of an electric potential (i.e. electric field). Plasma, on the other hand, describes the state of the gas. It is often called the fourth state of matter[11], but the term is closely tied to gas discharges.

A plasma is essentially a gas that consists of free ions and electrons and therefore is electrically conductive, but the overall state of the gas is charge neutral[12]. That is, there is an equal number of positive and negative charges throughout the gas[13]. Full decomposition of the gas into ions and electrons typically occurs at very high temperature, whereas at moderate and low (room) temperatures the plasma is *partially ionized.*

A discharge is, in contrast, simply the conduction of current across an insulating gas. But, in practice, a *self-sustained* discharge typically has a plasma (net neutral)

[11] Conceptually, as temperature increases a solid turns into a liquid, a liquid into a gas, and a gas into a plasma.

[12] The term 'plasma' was introduced in 1928 by Irving Langmuir, an American chemist and physicist and Nobel Prize winner.

[13] When the charges are not equal, we say there is *space charge*; if there are excess ions it is positive space charge and if there are excess electrons, it is negative space charge. This will be discussed more in chapter 9.

region[14]. The term gas discharge is used when describing non-equilibrium, low-temperature conditions and some high-temperature conditions where the degree of ionization is modest (like arcs), but is often used interchangeably with plasma. The term plasma is used exclusively when describing thermal plasmas at extreme temperatures, such as thermonuclear fusion. In this book, since we are concerned with laboratory plasmas, we will use the term gas discharge, or simply 'discharge', preferentially.

To summarize, plasmas are generally characterized by *how many* electrons there are (concentration) and *how energetic* they are (temperature). Figure 1.1 shows the phase space for different types of discharges and plasmas based on these two properties. Note that most non-equilibrium, low-temperature discharges have electron concentrations much smaller that the concentration of gas in a room at STP[15]. Thus, for a non-equilibrium, low-temperature discharge at atmospheric pressure, there is typically far less than 0.1% of the gas that is even ionized.

How are gas discharges generated? As we learned above, charged particles move under the influence of an electric or magnetic field, and thus it is logical to surmise that a gas discharge is generated when a sufficiently high electric or magnetic field is applied to a gas. However, as we will learn in part I, all it takes to create an ion is energy—whether that be thermal, light (photon), electrical, or even chemical.

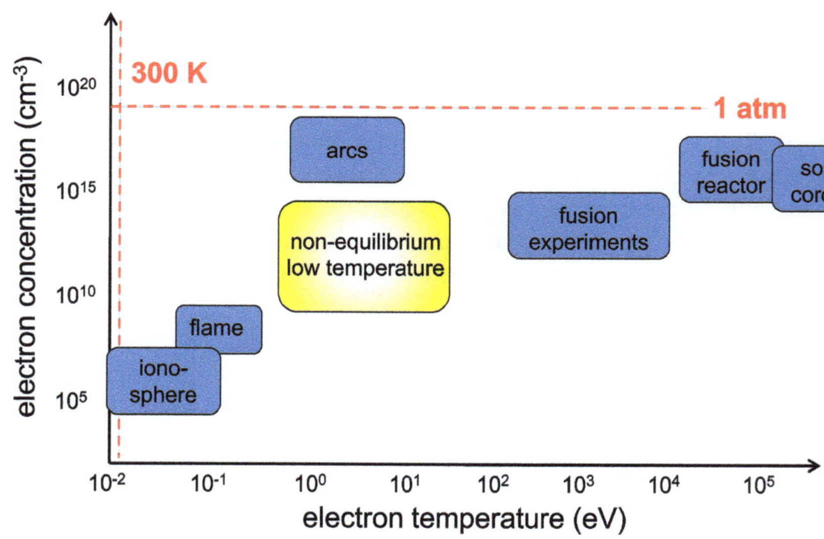

Figure 1.1. Phase space of different plasma regimes based on the electron concentration and electron energy (or temperature). Note that room temperature (300 K) and the gas concentration ($\sim 10^{19}$ cm^{-3}) at atmospheric pressure and room temperature are marked to provide some calibration. A typical non-equilibrium, low-temperature discharge has an electron concentration many, many orders of magnitude smaller that the gas concentration of atmospheric air and an electron temperature many orders of magnitude higher than room temperature.

[14] This will also be discussed more in chapter 9.

[15] STP = standard temperature and pressure.

Therefore, there are many different ways of generating discharges and plasmas beyond simply applying electric or magnetic fields, such as using lasers or microwaves. However, for the sake of simplicity, we will focus solely on electric fields. There are two basic strategies to use an electric field to generate a discharge.

Direct current (dc) discharges are generated by the application of a *static electric field* between two (or more) metal electrodes. The shape of the electrodes largely dictates the discharge mode, with glow, corona, and hollow cathode discharges being common examples.

Alternating current (ac) discharges are generated by the application of an unsteady, *periodic electric field*, and this can be between two metal electrodes but also electrodes covered in insulating, dielectric material. Perhaps the most common examples are radio frequency (RF) discharges (such as a capacitively coupled plasma) and dielectric barrier discharges (DBDs), which typically use sinusoidal ac electric fields. However, pulsed dc and pulsed DBD configurations, including both microsecond and nanosecond pulses, are also common configurations.

1.3 Summary

It is clear from this brief overview that the field of gas discharges and plasmas can be very complex, spanning many orders of magnitude in properties and characteristics. However, the basic science underlying the majority of them is the same, and this is certainly the case for non-equilibrium, low-temperature discharges. In the first two parts of this book, we will look at the underlying principles of the two main processes in discharges, ionization and ion transport, and appreciate the foundation that forms the field of plasma science and engineering. In the third part, we will apply these underlying principles to understand basic gas discharge concepts, including how they are formed, their basic structure, and basic properties, focusing on the simplest configuration of a DC glow discharge. The structure for each part is largely the same, leading with essential physics concepts and followed by more detailed discussions relevant to gas discharges, although part III necessarily looks a little different. For some of the topics covered here, detailed mathematics will be incorporated in an effort to make the physics clear, whereas in other cases, the detailed mathematics will be omitted for brevity. For all topics, the primary emphasis will be on providing the reader with the essential physical and chemical concepts that are needed for further study in the field of plasma science and engineering.

Part I

Ionization

IOP Publishing

Ionization and Ion Transport (Second Edition)

A primer for the study of gas discharges and plasmas

David B Go

Chapter 2

Foundations from gas dynamics

The term *ionization* broadly refers to the generation of free, gaseous charges such that the interstitial gas between two electrodes transforms from an *insulator* into a *conductor*. There are essentially two types of processes that are important in the creation of charges. The first is gaseous processes (or volume processes)—that is, the charges are created from the gas particles in the volume themselves. The second is surface (or, more precisely, electrode) processes—that is, the charges are created from the solid structure of the electrodes and injected into the gas. Generally, the charges of interest are electrons, but because electrons are often created during the same process as ions, this part of the book is called 'Ionization'. Both of these processes will be covered here, including some mechanisms that are generally not important, so that the reader can understand the dominant versus negligible processes. This part of the book is organized into four chapters. The first two establish the foundational concepts necessary to understand ionization. First, in chapter 2, basic concepts of gas dynamics and statistical mechanics are discussed to provide general descriptions of a gas. We then extend these concepts to specifically focus on electrons in the gas in chapter 3. We apply these fundamentals to understand gaseous ionization in chapter 4, and then finish our discussion with electrode processes in chapter 5.

As noted in the introduction, a discharge is an ionized gas, and therefore it is also a gas. In order to establish a firm physical foundation for the study of ionization, it is necessary to start with some of the basic principles that describe a gas generally, and then see how these are modified or different when describing a discharge.

First, we must ask ourselves what makes up a gas, and for the purposes of this discussion we will focus on pure gases[1] that are made up of either atoms (an 'atomic' gas) such as helium (He) or molecules (a 'molecular' gas) such as nitrogen (N_2).

[1] Obviously this oversimplifies things in the sense that the most common gas we as living creatures on Earth experience is air, which is comprised of nitrogen, oxygen, argon, helium, and any number of various pollutants and particles. Certainly air discharges are important, and for those interested in that particular area the book *Non Equilibrium Air Plasmas at Atmospheric Pressure*, edited by Becker *et al* [1] is recommended.

doi:10.1088/978-0-7503-3991-9ch2

Now, when we think about this gas, we can think about it in one of two ways. First, we can think about every single individual particle (atom or molecule) and how it moves and bounces around or we can think of the gas as this continuous medium where all the particles move together, much like we think of a solid. The first viewpoint is called a kinetic perspective, where we consider how to describe the behavior of a collection of individual particles, and the second viewpoint is a continuum perspective, where we view the gas as a single, continuous fluid. In terms of common engineering topics, the former is gas dynamics and the latter is classical fluid dynamics. Likewise, we can consider discharges from both the kinetic and continuum perspectives. This chapter will introduce essential concepts from gas dynamics and statistical mechanics that are important for appreciating the underlying processes that govern discharges[2]. Treating the discharges as a continuous fluid will be discussed in more depth in chapter 7.

2.1 The atom, the molecule, and excited states

As with all matter, discharges are made up of atoms and molecules, and the structure and behavior of these individual particles defines the structure and behavior of the discharge as a whole. Therefore, we will start at the very basic level of a single atom (or single molecule). Recall that an atom is made up of three primary sub-atomic particles. The nucleus is the center-of-mass of the atom and consists of neutrons and protons[3] and surrounding the nucleus are electrons[4]. Atoms generally have net neutral charge (i.e. the same number of electrons and protons) and thus there are no 'free' charges in a gas that are able to move if placed in an electric or magnetic field. However, in order for the gas to be conductive, the charges must be free to move, and this typically consists of either unattached electrons (free negative charge), atoms/molecules that have lost one or more electron (positive ions), and atoms/molecules that have gained one or more extra electrons (negative ions). One could say that electrons, therefore, are the most important fundamental particle in a discharge.

Atoms, molecules, and sub-atomic particles at this scale do not behave following classical physics and instead exhibit behavior described by quantum mechanics. That is, the main quantities of the particles such as energy and momentum are restricted to discrete, or quantized, values. This is in contrast to the macroscale, where these quantities are said to be continuous[5]. Since atoms and molecules are free to move, they, like all other matter, can have translational kinetic energy and momentum. It can be shown from the underlying quantum mechanics[6] that the translational energy for an atom[7] is

[2] There are many good texts for studying these topics, and one that is particularly useful is *Statistical Thermodynamics: Fundamentals and Applications* by Laurendeau [2].

[3] Neutrons: zero (0) charge, ~1.675×10^{-27} kg; protons: +1 charge, ~1.673×10^{-27} kg.

[4] Electrons: −1 charge, ~9.109×10^{-31} kg.

[5] For example, when we consider the speed of a car, we assume it can be of *any* value between 0 km h^{-1} and some large number, say 100 km h^{-1}.

[6] In short, Schrödinger's equation is solved, and the eigenvalues result in discrete energies.

[7] A similar expression can be derived for a molecule, but is adjusted to account for the center-of-mass.

$$E_{tr} = \frac{h^2}{8mV^{2/3}}(n_1^2 + n_2^2 + n_3^2), \tag{2.1}$$

where h is Planck's constant[8], m is the mass, and V is an effective volume. The three n parameters are called the *translational quantum numbers* and can be any positive integer, $n_i = 1, 2, 3, \ldots$. These quantum numbers are what give the energy levels discrete values and forbid energies except at specific values. What is interesting is that it can be shown that these levels are *so close* together that they look essentially continuous. However, this concept of a quantum number arises in other energy modes, and the impact of discreteness is much more noticeable.

While it is fairly obvious that atoms and molecules in a gas will have translational energy, they will also have additional energy states due to their internal structure as well as the interaction between sub-atomic particles (in an atom) or the interaction between atoms (in a molecule). Consider first the electrons. The electrons in an atom occupy orbitals that surround the nucleus of the atom, and the distribution of electrons in the atom is described by four *quantum numbers* as follows:

1. The *principal quantum number* represents the shell (i.e. main energy level) occupied by the electron and it is a positive integer, e.g. $n = 1, 2, 3, \ldots$. These shells are typically denoted by the upper case letters K, L, and M.
2. The *secondary quantum number* represents the orbital angular momentum (it is also called the orbital angular momentum quantum number or azimuthal quantum number) and reflects the type or shape of the orbital. These are denoted by lower cases letters s, p, d, f, g, h, and i. It depends on the value of the principle quantum number n and takes a value of $l = 0, 1, \ldots, n-1$.
3. The *magnetic quantum number* represents the number of possible orientations for each type of orbital and depends on the value of the secondary quantum number l, taking a value of $m = -l, \ldots, 0, \ldots, l$.
4. The *spin quantum number* represents the two possible orientations an electron can have in the presence of a magnetic field (spin up or spin down), and is either $s = -1/2$ or $+1/2$. Based on the Pauli exclusion principle, only electrons with opposite spin (opposite s) may occupy the same orbital.

A key concept here is that it is generally impossible for two electrons in an atom to have the same four values of these quantum numbers.

Now, an **excited state** is a quantum state of an atom that is above its ground state. That is, when all the electrons in an atom are in their lowest (minimum) energy state, it is in its ground state. However, if energy is input into the atom, the electrons will be energized into a different quantum configuration (a different set of quantum numbers), and this is called an excited state[9]. Importantly, when the electron relaxes

[8] $h = 4.135 \times 10^{-15}$ eV s is named after the German Max Planck, one of the most influential theoretical physicists in history, and whose work on quantized energy led to the 1918 Nobel Prize.

[9] A very simple picture is that if the atom absorbs energy from a laser, an electron might be excited from an inner shell, K, to an outer shell, L, and so forth.

back to the lower excited state, it will emit that gained excess energy as a photon. Most discharges 'glow' and this is because atoms and molecules that have been excited relax to lower energy levels and give off light[10]. In a very simplistic sense, we can visualize the excited states via an energy level diagram, as shown in figure 2.1.

Critically, for an atom to be excited, it must receive at least as much energy as it takes to put the electron into a higher energy level; the electron cannot occupy any energy between two energy levels. These are often called threshold energies. Additionally, at some critical energy the electron has sufficient energy to break away from the nucleus, becoming a 'free' electron and leaving behind a positively charged ion with one excess proton. This critical energy is called the **ionization energy** and depends on the type of atom or molecule[11]. Take, for example, helium gas. Helium has upwards of 49 excited states with the threshold energy to get into the first excited state being 19.82 eV and the ionization energy being 24.59 eV.[12] That is, there are 49 possible energy levels for electrons between 19.82 and 25.59 eV. In contrast, the nitrogen molecule N_2 has about 29 excited states with the first excited state at 6.725 eV and the ionization energy at 15.58 eV.[13]

Molecules are more complicated because there are additional energy states that correspond to the motion of the atoms of the molecule against each other. The atoms can vibrate against each other or the atoms can rotate about each other. Just like electrons, both the vibrational and rotational motion are governed by quantum mechanics. For simplicity, consider a diatomic molecule consisting of only two

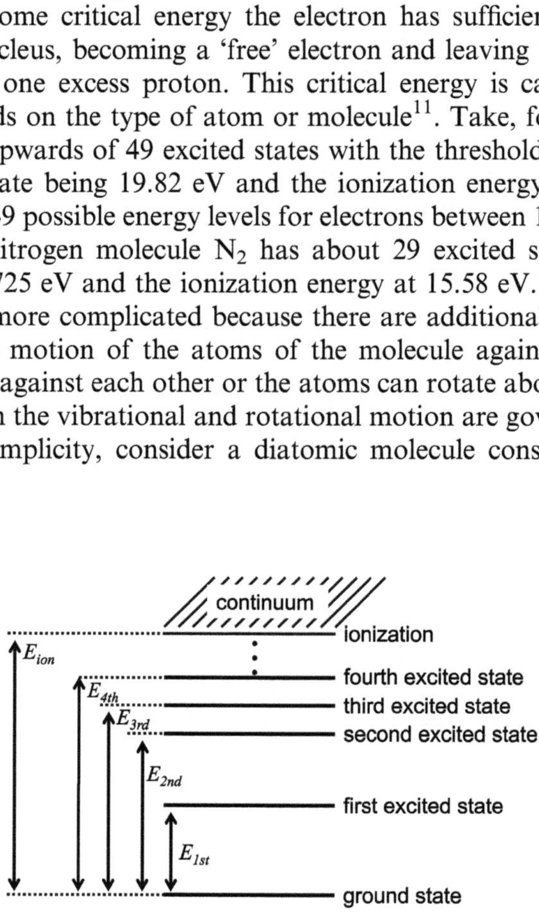

Figure 2.1. Illustrative energy level diagram for an atom.

[10] For example, a neon sign is a type of glowing discharge.

[11] Note that ionization energy is often presented as either kJ mol^{-1} or eV per atom, where the unit conversion is 1 kJ mol^{-1} = 0.010 364 eV per atom.

[12] Most resources with atomic data will have the energy levels and ionization energies of the elements. One such resource is *Reference Data on Atoms, Molecules, and Ions* by Radzig and Smirnov [3].

[13] If we consider that the particle is excited by a photon, 19.82 eV corresponds to ~62 nm or extreme ultraviolet (EUV) light, whereas the 6.725 eV corresponds to ~184 nm or vacuum ultraviolet (VUV) light.

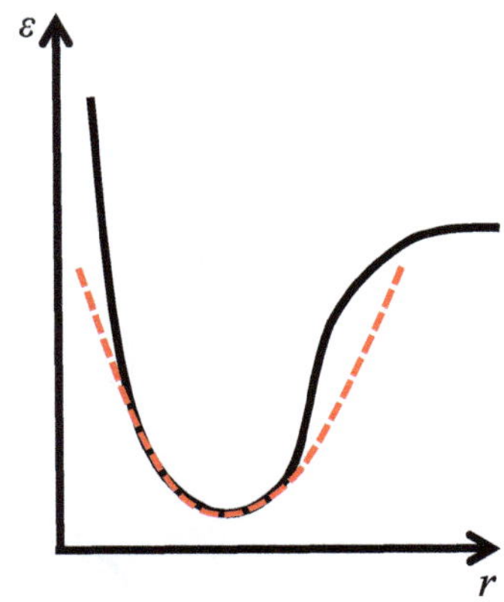

Figure 2.2. Illustrative diagram of the potential energy between two atoms in a molecule, with the red dashed line showing the parabola at the bottom of the well that leads to a harmonic oscillator approximation.

atoms, such as N_2. The potential energy due to the chemical bond between the two N atoms forms a well, as shown in figure 2.2. If you try to push the N atoms closer together (small r) the energy is great, and the reactive force pushes the atoms apart. Similarly, if you try to pull the atoms apart, the energy is great, and the reactive force tries to pull them back together. At equilibrium, the energy is minimized, and this is the bottom of the well.

This push–pull phenomenon is very similar to a spring, and the bottom of the potential well can be approximated by a simple parabola. In this way, the vibrations can be modeled using simple harmonic motion that can be described by a vibrational frequency ω. The lowest or ground vibrational energy level is called the *zero-point energy* and corresponds to

$$E_{\text{vib, 0}} = \frac{1}{2}\hbar\omega, \tag{2.2}$$

where $\hbar$ is the reduced Planck's constant[14]. The equally spaced vibrational energy levels then correspond to a *vibrational quantum number* v as

$$E_{\text{vib}} = \left(v + \frac{1}{2}\right)\hbar\omega, \tag{2.3}$$

where $v = 0, 1, 2, 3, \ldots$.

Molecular rotations are often described by considering the molecule to be a rigid rotor. Similar to vibrational energies, they can be described by a simple formula

$$E_{\text{rot}} = \frac{\hbar^2 J(J+1)}{2I}, \tag{2.4}$$

[14] The reduced Planck's constant is simply $\hbar = h/2\pi = 6.582\ 119\ 514 \times 10^{-16}$ eV s.

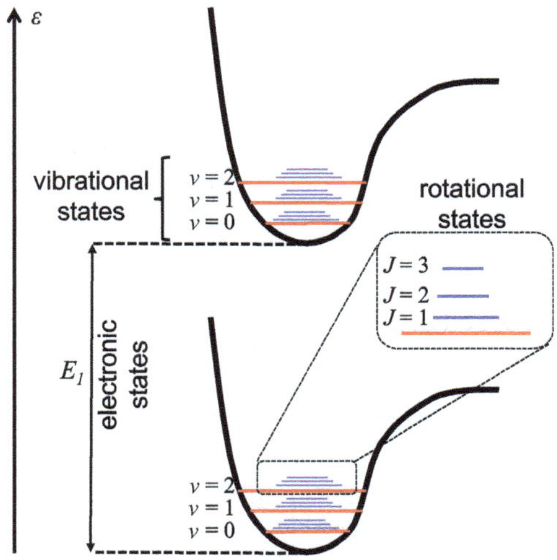

Figure 2.3. Illustrative energy level diagram for a molecule showing the electronic, vibrational (red), and rotational (blue) energy levels.

where I is the moment of inertia of the molecule and J is the *rotational quantum number*. Similar to v, it must be 0 or a positive integer, $J = 0, 1, 2, 3, \ldots$.

Vibrational and rotational energies are much smaller than electronic energies. For example, the first N_2 vibrational energy level is only ~0.29 eV and each subsequent energy level is incremented by 0.29 eV. The first rotational level is even smaller, ~1.45×10^{-3} eV. Contrast these to the 6.725 eV required to energize N_2 into its first electronic state. Furthermore, the rotational and vibrational energy levels exist in addition to the electronic energy levels, suggesting our simple sketch in figure 2.1 is actually too simple when we consider molecules. One way to think of it is that the energy levels are embedded—a molecule has several electronic energy levels, each electronic energy level has vibrational energy levels, and each vibrational energy level has rotational energy levels. We can thus redraw our picture of the energy levels for a molecule as shown in figure 2.3.

The key take-away is that if energy is input into a gas, such as thermally or via light, the atoms and molecules in the gas can be excited. Gas discharges input electrical energy into a gas to excite these particles; we will cover the exact mechanism in chapter 3. But before we get there, we need to understand a gas not just from a single particle perspective but also from a many particle perspective.

2.2 The statistics of a gas

Now that we have a clearer picture of the particles that make up a gas, we need to start to consider how we understand them from a collective perspective. Suppose we wanted to understand a system of N gas particles in a volume V[15]. To understand the system, we need to evaluate the position and velocity of each particle. If each particle is interacting with each other through some force[16], then we would need to solve

$$m_i \frac{d\vec{v}_i}{dt} = \sum_{j=1}^{N-1} \vec{F}_{ij}(\vec{r}_i, \vec{r}_j, t) \quad \text{for} \quad i = 1, 2, 3, \ldots, N.$$

This is just not possible even given the computing power of modern technology. Statistical mechanics helps us to understand the *microscopic* properties of the gas and relate them to the *macroscopic* properties of the gas by obtaining the **equilibrium distribution** of particles in the gas. Kinetic theory considers the transport of individual particles under non-equilibrium conditions in order to relate microscopic properties to macroscopic transport properties. In many cases, gas discharges are not at equilibrium, as we will see, and we will need to take a kinetic perspective when analyzing the transport and transport properties. However, before doing so, we will establish a basic foundation in statistical and kinetic concepts.

If we have a gas of N particles, each with their own energy ε, we can organize them into 'energy levels' each with some subset N_i of the total number of particles, as shown in figure 2.4. We call each energy *level* ε_i containing N_i particles a **macrostate**, and each macrostate consists of individual energy *states* called **microstates**. These microstates are based on the quantized energy that is a result of the quantum mechanics as discussed in section 2.1. The macrostates or energy levels therefore have subsets consisting of the quantum microstates. We call these microstates the degeneracy g_i of the macrostate ε_i. The number of microstates corresponds to the thermodynamic probability W of being in that macrostate—that is, we define **thermodynamic equilibrium** as the *most probable* macrostate.

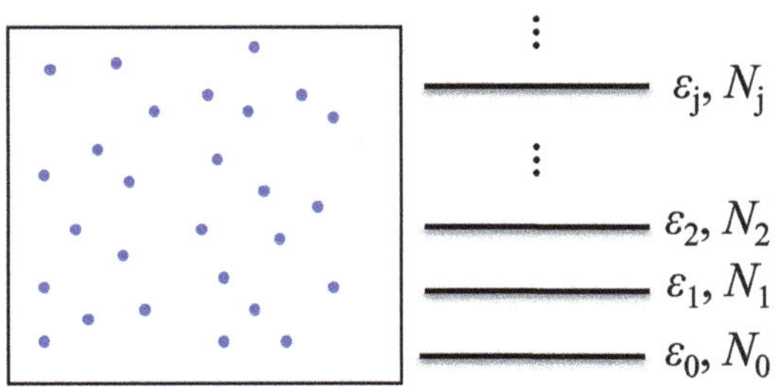

Figure 2.4. A gas with N particles split into different energy levels ε_i.

[15] For calibration, there are ~10^{16} gas particles in 1 mm^3 volume at standard temperature and pressure.

[16] For example, gravitational or Coulombic.

If a certain number of assumptions are made[17], then Maxwell–Boltzmann statistics[18] gives us the number of microstates per macrostate or the thermodynamic probability[19]

$$W = N! \prod_{i=0}^{\infty} \frac{g_i^{N_i}}{N_i}. \tag{2.5}$$

This thermodynamic probability leads to a number of important gas properties. Using Lagrange multipliers to convert the product to an exponential, it can be shown that the fraction of particles at a specific energy level, or, more eloquently, the *probability* of a sub-population of particles being at a specific energy level i is

$$\frac{N_i}{N} = \frac{\exp\left(\frac{-\varepsilon_i}{k_B T}\right)}{\sum_{i=0}^{\infty} g_i \exp\left(\frac{-\varepsilon_i}{k_B T}\right)}. \tag{2.6}$$

This, in effect, defines how the particles are distributed among the various energy levels. The summation in the denominator is called the molecular *partition function* Z, which is equivalently

$$Z = \sum_{i=0}^{\infty} g_i \exp\left(\frac{-\varepsilon_i}{k_B T}\right). \tag{2.7}$$

The probability of particles being in a specific energy level then formally becomes

$$\frac{N_i}{N} = \frac{\exp\left(\frac{-\varepsilon_i}{k_B T}\right)}{Z}. \tag{2.8}$$

This partition function is crucial. If we know the energy levels ε_i and their degeneracies g_i (number of microstates per energy level) for a gas, then we know how the particles in the gas occupy these energy levels. For example, the partition function associated with *translational* energy[20] Z_{tr} can be derived by applying quantum mechanics to the so-called 'particle-in-a-box' to obtain

$$Z_{tr} = \left(\frac{2\pi m k_B T}{h^2}\right)^{3/2} V. \tag{2.9}$$

---

[17] Assumption 1. The time-average for a thermodynamic variable is equivalent to the average over all possible microstates. Assumption 2. All microstates are equally probable. Assumption 3. The particles are independent.

[18] Named after James Clark Maxwell and Ludwig Eduard Boltzmann, whose tombstone bears his equation for entropy $S = k_B \log(W)$. Interestingly, they derived these formulae independently.

[19] Note that equivalent thermodynamic probabilities can be derived to describe electrons, called Fermi–Dirac statistics, which affect surface processes as discussed in chapter 5, and to describe phonons, called Bose–Einstein statistics.

[20] Note that there are partition functions associated with translational and electronic energies in an atom, as well as rotational and vibrational energies in molecules.

If the partition function Z is known because the energy levels ε_i and their degeneracies g_i are known, we can describe a number of important thermodynamic properties including the internal energy, entropy, Gibb's free energy, and so forth[21]. While equation (2.8) describes the probability of the particles in a gas being in any given energy state, it is difficult to connect how this distribution of energy states ultimately relates to properties that we can predict and, more importantly, measure. Kinetic theory allows us to extend this description and formalize it to determine, for example, how the complete ensemble of particles is distributed and ultimately, important gas properties.

2.3 Kinetics

In the continuum perspective, we treat a gas (or fluid or solid) as a uniform body that moves as a single continuous entity. A gas flow at 1 m s^{-1} implies the entire gas medium is moving at 1 m s^{-1}. However, at the particle level, each particle is moving at its own speed and undergoing collisions with its neighbors; therefore, each particle has a different velocity, and hence, energy. In order to understand the collective, we need to understand how they interact and apply appropriate statistics to arrive at characteristic properties of the whole.

2.3.1 Collision theory

As discussed in section 2.1, as energy is input into a gas, the particles of the gas can be excited into various energy states (translational, rotational, vibrational, and electronic). The question, however, is how is the gas excited?

Let us consider a 'box' of 'marbles' as an analogy; we might consider that a 'volume' of 'gas particles' might behave in a similar way. Just as if you shake the box, the marbles will bounce around and collide with each other, the particles in gas are constantly bouncing around and colliding with each other. As we add energy by shaking the box, the marbles will move around more, with the key concept that energy is *input* into the system through *collisions with the wall.* That is, the initial shaking of the box has the walls of the box hit the marbles at the edges and then these edge marbles collide with interior marbles, transferring the kinetic energy from the box to all the marbles in the box. Similarly, if we were to add heat to a volume of gas, say by heating one of the surfaces, the atoms/molecules *in* the surface vibrate within the bonds of the material and exchange greater energy with the gas particles that collide with the surface. These gas particles now have greater energy and collide with interior particles, exchanging momentum and energy. In this way, energy *diffuses* into the gas from the surface. Clearly then, collisions between particles dictates how the particles behave as a collective (or ensemble). While we will discuss later, in chapter 3, how these collisions affect ionization and, in chapter 6, how these collisions affect transport, for now we just want to understand the basics of collision theory.

[21] One of the most impressive aspects of statistical mechanics is how these particle-scale descriptions can be used to determine macroscopic properties. More details can be found in a text such as [2].

One easy way to visualize collisions is by considering a collection of very fast small particles of radius r_2 moving through a region full of large, slow particles of radius r_1—an approach often called 'beam' theory because we have a beam of small particles moving through a 'quasi-stationary' group of large particles, as illustrated in figure 2.5[22]. As this beam of small r_2 particles interacts with the r_1 particles, the r_2 particles will spread out, changing their local density. If we assume these particles are solid spheres, like billiard balls, then a collision occurs when the distance between the beam particles and stationary particles is less than $r_1 + r_2$. This is shown schematically in figure 2.6. We can define the **collision cross-section** σ as the circular plane carved out by this distance $r_1 + r_2$ or[23]

$$\sigma = \pi(r_1 + r_2)^2. \tag{2.10}$$

By this definition, the collision cross-section is an *area* and thus has units of m^2.[24] So one way to think of it is that the collision cross-section is the area *blocked* by particle

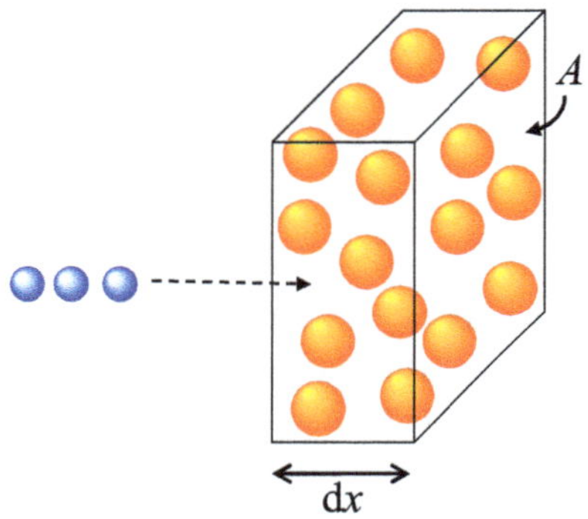

Figure 2.5. Conceptual schematic of beam of particles passing through a volume of stationary particles.

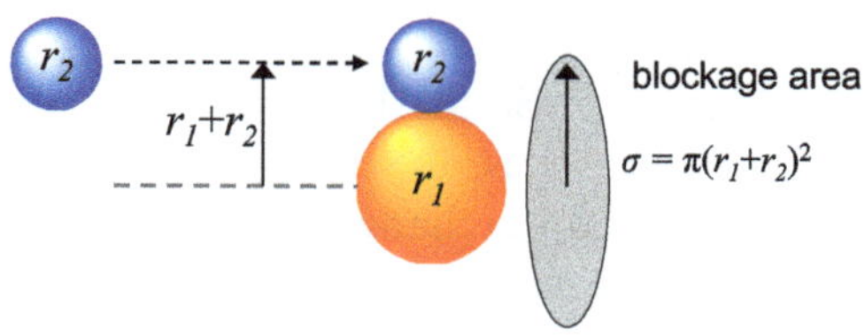

Figure 2.6. Schematic of the blockage area in the hard-sphere approximation.

[22] A similar concept also applies to light scattering and thus a light beam.

[23] Note that this is called the hard-sphere approximation. Real interactions between particles have much more complex collision cross-sections that account for the atomic interactions between the two particles. We will cover that in more depth in chapter 3.

[24] Although it is also commonly reported in cm^2.

1. As we will see in chapter 3, the collision cross-section is a critical fundamental concept of gas discharges. But for now, it is easier to think of it phenomenologically as being related to the *probability* that two particles will collide. Clearly, if the collision cross-section is large, then the probability that two particles will collide is high—two large particles having a greater chance of hitting each other than two small particles.

One essential concept that depends on the collision cross-section is the **mean free path**. The mean free path l is the average distance traveled by a particle in between collisions with other particles, as shown in figure 2.7. While one could define the mean free path based on simply averaging the distance between every collision, this would require actually calculating or measuring each individual collision. However, we can determine this average more easily by considering the probability of a collision.

Consider again the beam of particles entering a control volume of size dx and cross-sectional area A containing stationary particles, as shown in figure 2.5. The stationary particles block the path of the beam particles, and the total blockage area is given by the collision cross-section for each particle multiplied by the number of stationary particles in the area or[25]

$$A_{\text{block}} = (N_1 \cdot A \cdot dx)\sigma. \tag{2.11}$$

The probability that one of the incoming beam particles will collide with one of the stationary particles is simply the ratio of the blockage to the total area or

$$P = \frac{A_{\text{block}}}{A} = \frac{(N_1 \cdot A \cdot dx)\sigma}{A} = N_1 \sigma dx. \tag{2.12}$$

Based on this probability, it can be shown[26] that the mean free path is

$$\bar{l} = \frac{1}{\sigma N_1}. \tag{2.13}$$

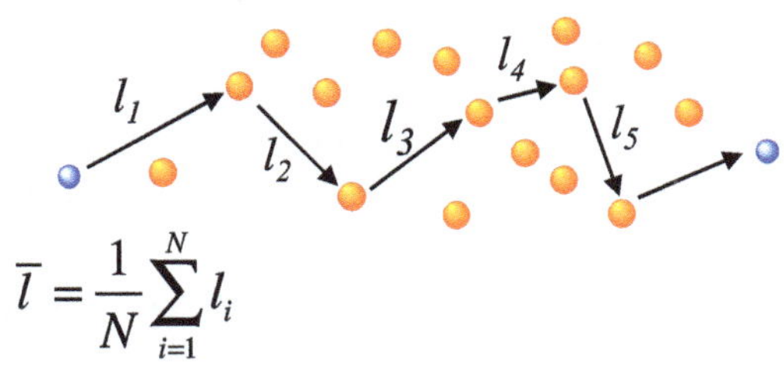

Figure 2.7. Conceptual schematic of mean free paths.

[25] Recall that N_1 is the number *density* so it needs to be multiplied by the volume of the control volume to give the total number of particles in the control volume.

[26] Formal derivations of the mean free path can be found in books such as [4] or even on Wikipedia at https://en.wikipedia.org/wiki/mean_free_path.

Equation (2.13) reveals some very important information. For instance, it is clear that the mean free path is *inversely* related to the collision cross-section, and thus the size of the particles that are in the gas. For example, using the hard-sphere approximation in equation (2.10) for the collision cross-section, the mean free path becomes

$$\bar{l} = \frac{1}{N_1\pi(r_1 + r_2)^2}. \tag{2.14}$$

Again, this makes sense phenomenologically; one would expect that the mean free path would be smaller if there were larger particles involved. Additionally, equation (2.13) shows that the mean free path is inversely related to the number density of stationary particles. Again, this makes sense as one would expect more collisions the more particles there are with which to collide.

If we invoke the *ideal gas law*, we can relate N to the gas pressure p and temperature T via $P = Nk_bT$. Thus equation (2.13) becomes

$$\bar{l} = \frac{k_bT}{\sigma P}. \tag{2.15}$$

Here we see that the mean free path is inversely related to the gas pressure.

2.3.2 Equilibrium behavior

One of the most important aspects of collisions is that they lead to the equilibration of a gas. That is, collisions serve to slow down fast particles, as they lose energy in collisions, and speed up the slow particles, as they gain energy from the fast particles until the entire ensemble reaches some equilibrium. While we would not expect all the particles to have exactly the same properties—it is rational that there will always be some variation in the velocity, speed, and energies of individual particles—we do anticipate that these will be distributed about some mean value. That is, we anticipate that there will be probability density functions (PDFs)[27] that describe the relative probability of particles in the entire population having a specific velocity, speed, or energy. We call these the distribution functions for the kinetic properties, and from the partition functions described in section 2.2, it is possible to determine the distributions *at equilibrium* for these properties; these are called the *Maxwell–Boltzmann distributions.*

The **Maxwell–Boltzmann** distribution for the particles' momentum $\vec{p}$ is

$$f(\vec{p}) = \left(\frac{1}{2\pi mk_BT}\right)^{3/2} \exp\left(-\frac{p_x^2 + p_y^2 + p_z^2}{2mk_BT}\right), \tag{2.16}$$

[27] What is a *probability density function* (PDF)? It is a function that describes the likelihood that some variable x will take a value. It is typically a continuous distribution written as $f(x)$ and has units of x^{-1}. The *probability distribution function* is the integral of the PDF—that is, the probability that x is less than or equal to some chosen value x^* or $Pr[x \leqslant x^*] = \int_{-\infty}^{x^*} f(x)dx$. One essential property of PDFs is that $\int_{-\infty}^{\infty} f(x)dx = 1$. In other words, the probability that x exists between $-\infty$ and $+\infty$ must be 1—that is, x must exist.

where m is the mass of particles. Since momentum is related to velocity by $\vec{p} = m\vec{v}$, a Maxwell–Boltzmann velocity PDF can be determined by the relation $f(\vec{v})dv_x dv_y dv_z = f(\vec{p})dp_x dp_y dp_z$, to arrive at

$$f(\vec{v}) = \left(\frac{m}{2\pi k_B T}\right)^{3/2} \exp\left(-\frac{m(v_x^2 + v_y^2 + v_z^2)}{2k_B T}\right). \tag{2.17}$$

Perhaps more interesting is the corresponding distribution for a single direction, such as v_x, which is simply

$$f(v_x) = \left(\frac{m}{2\pi k_B T}\right)^{1/2} \exp\left(-\frac{mv_x^2}{2k_B T}\right). \tag{2.18}$$

This PDF is plotted in figure 2.8(a). At equilibrium, v_x is equally distributed in both the positive and negative directions. This is a normal or Gaussian[28] distribution, and it is clear from the figure that the mean x-velocity is 0. We can show this formally by taking the first central moment of the PDF or

$$\overline{v_x} = \int_{-\infty}^{\infty} v_x f(v_x) dv_x = \left(\frac{k_B T}{2\pi m}\right)^{1/2} \int_{-\infty}^{\infty} \exp\left(-\frac{mv_x^2}{2k_B T}\right) d\left(-\frac{mv_x^2}{2k_B T}\right) = 0. \tag{2.19}$$

Note the physical significance of this PDF. This implies that in a group of particles at equilibrium, any individual particle has as much likelihood of moving in the positive direction as the negative direction, such that their mean *directional* motion is zero. This matches conceptually with our vision of a room of gas particles that is nominally at 'rest' even though the individual particles are all moving.

The corresponding Maxwell–Boltzmann speed PDF can be derived by taking $v = \sqrt{v_x^2 + v_y^2 + v_z^2}$ and is

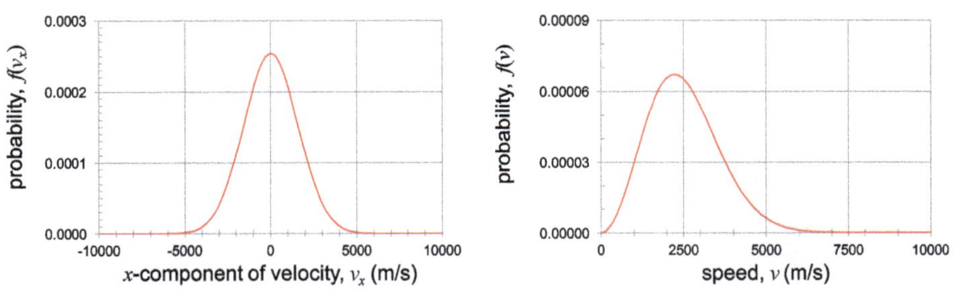

Figure 2.8. (a) The Gaussian distribution for the x-component of velocity for molecular hydrogen at 300 K. (b) The equivalent Maxwell–Boltzmann distribution for speed.

[28] Named after Johann Carl Friedrich Gauss, a German mathematician and physicist.

$$f(v) = \frac{4}{\sqrt{\pi}}\left(\frac{m}{2k_{\rm B}T}\right)^{3/2} v^2 \exp\left(-\frac{mv^2}{2k_{\rm B}T}\right). \tag{2.20}$$

This distribution is shown in figure 2.8(b). Note, unlike the PDF for v_x, the PDF for v is not centered around zero since this would imply that the mean speed is exactly 0 and thus no motion at all. Similarly, there is a long tail to the distribution that shows that a very small population of the particles in the ensemble have extremely high speeds. The mean speed can be determined similarly to the mean x-velocity and is

$$\bar{v} = \left(\frac{8k_{\rm B}T}{\pi m}\right)^{1/2}. \tag{2.21}$$

Additionally, a root mean square (RMS) speed can be extracted from the PDF as

$$v_{\rm RMS} = \sqrt{\left(\overline{\vec{v}^2}\right)} = \left(\frac{3k_{\rm B}T}{m}\right)^{1/2}. \tag{2.22}$$

Notice that the RMS speed is greater than the mean speed, $v_{\rm RMS} > \bar{v}$.

The Maxwell–Boltzmann energy distribution can be derived by relating the kinetic energy ε to the speed v by $\varepsilon = \frac{1}{2}mv^2$ arriving at

$$f(\varepsilon) = \frac{2}{\sqrt{\pi}}\left(\frac{1}{k_{\rm B}T}\right)^{3/2} \varepsilon^{1/2} \exp\left(-\frac{\varepsilon}{k_{\rm B}T}\right). \tag{2.23}$$

Like the speed distribution, the energy distribution is not centered around zero since $\varepsilon = 0$ implies no motion, and there is also a high-energy tail as shown in figure 2.9. As we will learn, the particles in gas discharges can also be described by distribution functions and these 'high-energy' tails are particularly important.

The mean energy of the ensemble can be derived two ways. The more rigorous way is via

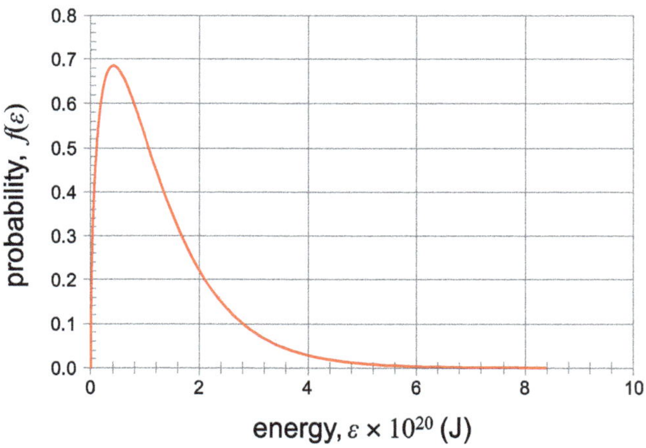

Figure 2.9. The Maxwell–Boltzmann energy distribution for molecular hydrogen at 300 K.

$$\bar{\varepsilon} = \int_0^\infty \varepsilon f(\varepsilon) d\varepsilon = \frac{3}{2} k_B T. \tag{2.24}$$

However, thinking about it more simply, the mean energy is simply the energy associated with the mean of the speed or

$$\bar{\varepsilon} = \frac{1}{2} m \bar{v}^2 = \frac{1}{2} m v_{\mathrm{RMS}}^2 = \frac{1}{2} m \left(\frac{3 k_B T}{m} \right) = \frac{3}{2} k_B T. \tag{2.25}$$

Not surprisingly, we arrive at the same value.

The key take-away from this relationship for the mean energy is that it is directly related to the temperature of the gas. In fact, this is essentially *what temperature is*—an expression for the average energy across a collection of particles—and for an equilibrium gas, the temperature is *defined* by the Maxwell–Boltzmann distribution. Under non-equilibrium conditions, if we consider just those particles energized into rotational states, we can find an equivalent rotational temperature for how those particles are distributed among rotational energy levels, and similarly vibrational and electronic temperatures as well[29]. In gas discharges, we use this concept to define the *temperature* of the various particles in the discharge, namely electrons, ions, and neutral particles, and these *temperatures* characterize how gas discharges are non-equilibrium, as discussed in chapter 1.

2.4 Summary

The purpose of this chapter was to give an overview of the basic gas dynamics and statistical mechanics concepts that are applicable to any system and will be invoked here for the study of gas discharges. There are a few primary points that were discussed:

- Atoms and molecules in a gas have quantized energy levels (translational, vibrational, rotational, and electronic) and can be excited out of the ground state (lowest energy level) into these excited states. However, after a short period of time, less than milliseconds, they will relax back to the ground state, releasing a photon.
- For a collection of atoms or molecules in a gas, each atom/molecule may occupy a different energy state for a given energy level. We can describe how the atoms/molecules are distributed among these various energy levels and states using statistical mechanics.
- In a collection of particles in a gas, the particles are all constantly moving and collide with each other; it is these collisions that distribute energy throughout the gas.
- The mean free path l is an important quantity that characterizes the average distance traveled by a particle between collisions.

[29] Note that this is an approach to defining temperature based on kinetic concepts, and it is a qualitative simplification at that. There are other approaches based on thermodynamics concepts.

- The equilibrium probability distribution functions for particles (speed, velocity, and energy) can be described using statistical approaches and are called the Maxwell–Boltzmann distributions.

Now it is critical to understand that the Maxwell–Boltzmann expressions reflect *equilibrium* conditions. In this case, the distribution of the particle energy states (translational, rotational, vibrational, electronic) are considered to be in equilibrium, and Maxwell–Boltzmann statistics applies. However, as we will learn in the ensuing chapters, gas discharges input electrical energy to perturb the gas into excited states—including ionized states—and furthermore, perturb the equilibrium distribution functions. Hence, in a gas discharge, the translational temperature is different from the rotational, vibrational, and electronic temperatures [5]. These concepts are not only important for producing ions, but also for driving the chemistry that makes gas discharges so useful.

Finally, it is important to note that while this section focused on basic concepts of statistical mechanics and kinetic theory, the field itself is remarkably rich in physics. One of the beautiful features of statistical mechanics is that using these basic concepts (i.e. the partition function), we can derive important thermodynamic properties such as entropy, internal energy, and Gibbs free energy *from first principles* as well as important relationships such as the ideal gas law *from first principles*. The elegance of this is quite profound. Similarly, kinetic theory can be used to derive *transport* properties of the gas, such as the viscosity and thermal conductivity, *from first principles*, which will be discussed in chapter 6. While this text will touch on some of these concepts as they relate to gas discharges, many of them will not be discussed, and so the reader is strongly encouraged to study statistical thermodynamics texts [2] for more details.

References

[1] Becker K H, Kogelschatz U, Schoenbach K H and Barker R J 2005 *Non Equilibrium Air Plasmas at Atmospheric Pressure* (Philadelphia, PA: Institute of Physics Publishing)

[2] Laurendeau N M 2005 *Statistical Thermodynamics: Fundamentals and Applications* (New York: Cambridge University Press)

[3] Radzig A A and Smirnov B M 2012 *Reference Data on Atoms, Molecules, and Ions* vol 31 (New York: Springer)

[4] Nassar E 1971 *Fundamentals of Gaseous Ionization and Plasma Electronics* (New York: Wiley)

[5] Fridman A and Kennedy L A 2011 *Plasma Physics Engineering* vol 2 (Boca Raton, FL: CRC Press)

IOP Publishing

Ionization and Ion Transport (Second Edition)
A primer for the study of gas discharges and plasmas
David B Go

Chapter 3

Elementary electron behavior

In chapter 2 we considered a gas as a set of atoms and molecules, but we have yet to discuss the charged particles that make a gas discharge conductive—those charges formed through ionization processes. The most important of these, of course, is the electron, and not just because it has charge. In fact, electrons are the most important particle in a gas discharge because they drive *the majority of all major processes*. How do they do this? As we discussed in chapter 2, particles in a gas collide and electrons are no different—they also collide with gas particles. It is these electron collisions that are so essential to gas discharges, and this chapter focuses on fundamental elementary concepts behind electron behavior.

3.1 Electron-driven gaseous reactions

The first question to tackle is what are the major electron processes in a gas. Consider a gas that also has some naturally occurring, free electrons[1]. When an electric potential (voltage) is applied across the gas, the free electrons will be accelerated by the resulting electric field, as discussed in the Introduction. Along the way, these electrons collide with the interstitial gas atoms or molecules, which we will generically refer to as M. These collisions represent chemical reactions, and they can be broadly classified into three groups (illustrated in figure 3.1):

Elastic: $e^{-}+M \rightarrow e^{-}+M$.

Excitation: $e^{-}+M \rightarrow e^{-}+M^{*}$ iff $\varepsilon_e \geqslant E_{M^*}$.

Ionization: $e^{-}+M \rightarrow 2e^{-}+M^{+}$ iff $\varepsilon_e \geqslant E_{\text{ion}}$.

Note that for the excitation and ionization reactions, the kinetic energy of the incident electron ε_e *must be* greater than or equal to the threshold energy required for excitation E_{M^*}, whether that be to a higher rotational, vibrational, or electronic

[1] Note that there is sufficient background radiation in the atmosphere creating electrons such that the free electron concentration is $\sim 10^2$ cm^{-3} at standard temperature and pressure (STP); this can be compared to the gas number density at standard pressure and temperature of $\sim 10^{19}$ cm^{-3} [1].

doi:10.1088/978-0-7503-3991-9ch3

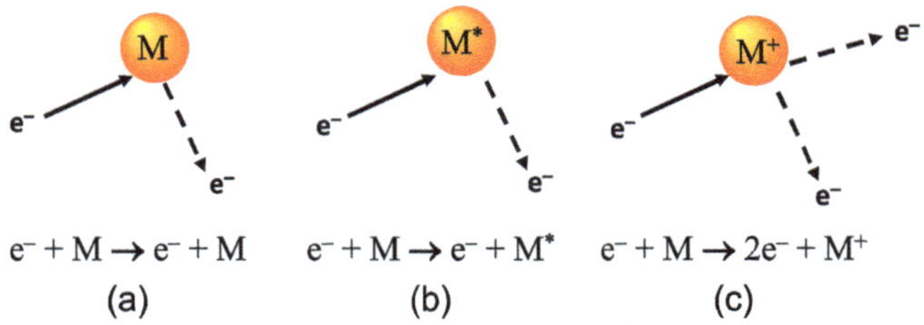

Figure 3.1. Basic reactions between a free electron and neutral molecule: (a) elastic collision, (b) excitation reaction, and (c) ionization reaction producing a second free electron.

state, or the energy required for ionization E_{ion}, respectively. Because some of the kinetic energy of the incident electron goes to exciting/ionizing the target atom, these are **inelastic** collisions as compared to the first reaction which is an **elastic** collision.

One often concludes that the reaction that is most interesting in the context of a discharge is the ionization reaction because it creates a second electron (plus a positive ion) and therefore is the charge producing reaction. However, all of these reactions are essential to the formation of a discharge as well as in various applications of gas discharges. Each of these is considered in a bit more detail below.

3.1.1 Elastic collisions

The elastic collision is the one that is easiest to conceptualize. The electron and molecule collide like two billiard balls, and all the energy of the collision is maintained (for the most part) as kinetic energy. That is, kinetic energy is conserved, and the collision is called elastic. In these binary (two-body) collisions, the electron is moving much faster than the gas particle, and therefore the gas particle is typically considered at rest relative to the electron. The electron collides with the gas particle and caroms off at a new trajectory. These are the predominant (i.e. most common) type of electron–particle collisions in a gas discharge and can occur for any kinetic energy of the incident electron ε_e[2].

3.1.2 Excitation collisions

As discussed in section 2.1, atoms and molecules have a variety of quantized rotational, vibrational, and electronic states. When a free electron is accelerated to sufficient energy, the kinetic energy of the electron can be transferred to the neutral particle, raising it to an excited state. Therefore, kinetic energy is *not* conserved in this collision, and the collision is called inelastic. Typically, an excitation collision/reaction requires an electron kinetic energy of a few to many electronvolts, depending on the excited state. For example, for molecules, rotational excitation only requires ~10^{-3}–10^{-2} eV, whereas vibrational excitation requires ~10^{-1}–10^{0} eV. Electronic excitation, on the other hand, requires at least ~5–20 eV for most common gases. As gas particles have a large number of possible excited states, and

[2] This, of course, is not exactly true, as we learned in section 2.1 that even translational energy is quantized. But the energy levels are so closely spaced that translational energy is essentially a continuum.

each has a different threshold energy, the kinetic energy of the incoming electron ε_e must exceed the threshold energy of an individual excitation reaction E_{M^*} for that specific excitation to be possible. The electron subsequently loses the threshold energy required for the reaction, such that after the collision the electron kinetic energy is

$$\varepsilon_e' = \varepsilon_e - E_{M^*}. \tag{3.1}$$

The reason a discharge glows is because once a neutral is excited, it will relax back to its ground state. In doing so, it releases that energy difference in the form of a photon, and, if the energy gap coincides with a visible wavelength, the photon is visible. For this reason, the color of a discharge depends of the type of gas used, because the wavelength of emitted light depends on the energy levels available to the atoms or molecules in the gas. (For example, a discharge in neon will be bright orange, while a discharge in argon will be more purple and blue.) Further, some of these states are more stable than others. Therefore, while some states relax quickly back to ground others stay in their excited (energized) state, and these are called **metastables**[3].

3.1.3 Ionization collisions

At sufficient kinetic energy, a free electron may actually cause an electron to detach from the neutral particle resulting in a second free electron and leaving behind a positive ion. The threshold energy required for this reaction is typically around 10–20 eV and is called the ionization energy E_{ion}. Therefore the incident free electron must have a kinetic energy of $\varepsilon_e \geqslant E_{\text{ion}}$ to create an ion. This is called **electron-impact ionization**, and it is the dominant way that free charges (electrons and ions) are produced in a gas discharge. This will be discussed in more detail in chapter 4.

Although excited states are not charge carriers themselves, in terms of the physics of the discharge, they play a crucial role in ionization because each time a free electron creates an excited neutral it *loses some kinetic energy* (the energy is spent exciting the neutral to a higher energy state) as shown in equation (3.1). As such, they act as an *energy sink*, delaying ionization reactions from occurring. Therefore, in order to accurately capture the ionization in a discharge, it is also necessary to understand the impact that excitation reactions play in modulating the energy of the electrons.

Finally, it should be noted that the three reactions discussed above are not the complete story. Other reactions, such as three-body reactions, electron-induced dissociation, and even double ionization, are also possible, though they only play significant roles under very specific regimes (such as at elevated pressures). The most important of these other reactions is *electron attachment*, where a free electron attaches to a neutral particle creating a *negatively charged atom or molecule*. Like

[3] From an application perspective, these excited and metastable states are very important because they are very chemically reactive. Many applications that utilize gas discharges, in fact, treat them as no more than gas-phase chemical reactors producing an abundance of reactive species.

other inelastic collisions, it also requires that the electron be at least at some threshold energy corresponding to the ground state of the resulting negative ion.

Attachment: $e^{-}+M \rightarrow M^{-}$ iff $\varepsilon_e \geqslant E_{\text{attachment}}$.

This is relevant specifically in electronegative gases, including oxygen (O_2) and sulfur hexafluoride (SF_6), and electron attachment can act as an electron sink, although the negative ions are still charge carriers producing current.

3.2 Electron–neutral collisions and the collision cross-section

3.2.1 The nature of electron collisions

In general, the nature of electron–neutral collisions is probabilistic. That is, even if an electron has sufficient energy to create an excited state or ion, there is only a small probability that this may occur. These probabilities are determined by the *collision cross-section* σ, which was first introduced in chapter 2. However, whereas the hard sphere model presented in equation (2.10) depends on only the *size* of the particles that are colliding, it should be clear that for electron–neutral collisions, they will also depend on the *kinetic energy* ε_e of the incident electron. That is, the type of collision is energy-dependent. At low incident electron kinetic energies (<0.01 eV), only elastic collisions are possible, at moderate energies (0.1–10 eV) elastic and excitation collisions can occur, and at high energies (>10 eV) elastic, excitation, and ionization reactions are all possible. Of course, the exact energy range is different for each gas and is determined by the threshold energies for excitation reactions into the different excited states of the gas and the ionization energy of the gas.

As we reformulate our understanding of the collision cross-section, there are two aspects we can intuit from what we now know. The first is that each *type* of electron–neutral collision, whether it be elastic, one of the many excitation reactions, or ionization, will have its own collision cross-section, i.e. its own reaction probability. Second, is that each of those collision cross-sections *must* have some dependence on the kinetic energy of the electron. How is this second point self-evident? Because excitation and ionization collisions cannot occur below a threshold energy, so at the very least, the probability must be zero for some electron energies and non-zero for other electron energies. The questions now are: 1. How is the collision cross-section different for different types of collisions? and 2. Why and how do they depend on energy? In this section, we will go through a more rigorous description of the collision cross-section in order to help answer these questions[4].

3.2.2 Differential collision cross-section

Let us again consider a particle (labeled as 2) moving at a velocity of $\vec{v_2}$ approaching a stationary particle (labeled as 1), as shown in figure 3.2. The closest distance of approach between particle 2 and particle 1 is what we call b, the **impact parameter**, and after the collision, particle 2 will deflect by an angle θ, which we call the **scattering angle**.

[4] For an even more detailed discussion, [7] is a good resource.

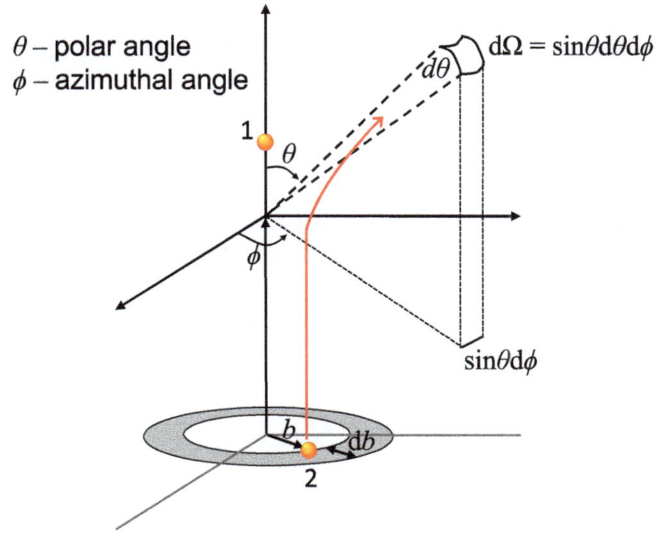

Figure 3.2. Schematic of the scattering of particle 2 due to a collision with stationary particle 1.

Now if we consider not a single approaching particle, but a beam of incoming particles, we can define the particle flux as $J_2 = N_2 v_2$. Importantly, conservation of particles requires that the number of *incoming* particles through the differential annulus of radius db must match the number of *outgoing* particles scattered through a differential solid angle $d\Omega = \sin\theta d\theta d\phi$, where θ is the polar angle and ϕ is the azimuthal angle. Another way of stating this is that all particles entering through a differential area $2\pi b db$ must leave through a differential angle of $\sin\theta d\theta d\phi$. However, this is not strictly correct because the differential angle has units of steradians, not area. Therefore, when we write this conservation we need to introduce a constant of proportionality I as

$$J_2(2\pi b db) = -J_2(I\cdot\sin\theta d\theta d\phi). \tag{3.2}$$

Note that the minus sign is because an increase in b leads to a decrease in θ.[5] We *define* the constant of proportionality I as the **differential scattering cross-section**, and this can be expressed mathematically as

$$I(\varepsilon, \theta) = \frac{b}{\sin\theta}\left|\frac{db}{d\theta}\right|. \tag{3.3}$$

The differential scattering cross-section is a function of the energy of the incident particle ε and scattering angle θ of the collision. The dependence on angle is readily apparent based on the definition in equation (3.3), but the nature (or origin) of the energy dependence is less clear. Also note that in order for equation (3.2) to be balanced, $I(\varepsilon,\theta)$ must have units of area.

[5] The further away the incident particle (larger b), the smaller the scattering angle θ.

We can now formally define the collision cross-section σ as the **total scattering cross-section** and equal to the integration of the differential scattering cross-section over the solid angle or

$$\sigma(\varepsilon) = \int_0^{2\pi} \int_0^{\pi} I(\varepsilon, \theta) \sin\theta d\theta d\phi = 2\pi \int_0^{\pi} I(\varepsilon, \theta) \sin\theta d\theta, \tag{3.4}$$

again with units of area. We can also define a *momentum transfer cross-section* by accounting for the fraction of momentum loss by the incident particle $1 - \cos\theta$ as

$$\sigma_m(\varepsilon) = 2\pi \int_0^{\pi} (1 - \cos\theta) I(\varepsilon, \theta) \sin\theta d\theta. \tag{3.5}$$

3.2.3 Elementary collision dynamics

It should be clear that σ is a function of energy because I is a function of energy. In order to determine $I(\varepsilon,\theta)$, it is necessary to determine a relationship between the impact parameter b and the scattering angle θ for equation (3.3). While doing so in the laboratory frame of reference is possible, it is more convenient to use a center-of-mass frame of reference. Consider figure 3.3(a) where particle 2 at velocity $\vec{v}_2$ is approaching particle 1 at velocity $\vec{v}_1$ at some impact distance b and after collision they both scatter at angles θ_2 and θ_1 relative to the axis of approach, respectively, with post-impact velocities $\vec{v_2'}$ and $\vec{v_1'}$. In a center-of-mass frame of reference, an 'imaginary' particle of reduced mass m_R approaching at a relative velocity $\vec{v}_R$ scatters against a fixed center-of-mass through angle Θ with a post-impact velocity of $\vec{v_R'}$.

If these particles are interacting, as all real particles do, then there is some equal and opposite force between them; particle 1 exerts a force $\vec{F}_{1,2}$ on particle 2 and particle 2 exerts a force $\vec{F}_{2,1}$ on particle 1. It can be shown that the equation of motion of the fictitious particle m_R is

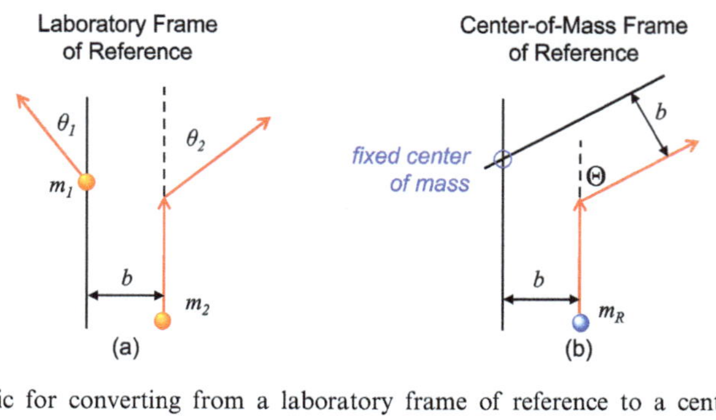

Figure 3.3. Schematic for converting from a laboratory frame of reference to a center-of-mass frame of reference.

$$m_R \frac{d\vec{v}_R}{dt} = \vec{F}_{2,1}, \tag{3.6}$$

However, as we recall from chapter 1, *force* is negative the gradient of the *potential* energy, $\vec{F} = -\vec{\nabla} U$. The critical concept here is that the collision will depend on the force of the interaction between the two particles and thus on the potential energy between the two particles. Importantly, all forms of potential energy (gravitational, elastic, electrostatic, etc) depend on the distance between the two objects. To illustrate why this is important, let us assume that the potential energy decreases as the distance r between the particles increases, or that

$$U(r) = \frac{C}{r^k}, \tag{3.7}$$

where k is some integer specific to the nature of potential energy and C is a constant of proportionality.

Now consider a fictional particle traveling relative to a fixed center-of-mass, as shown in figure 3.4, where the particle distance from the fixed center can be described as

$$\vec{r}(t) = \sqrt{b^2 + (v_R t)^2} \tag{3.8}$$

if we assume the scattering angle is very small[6]. Utilizing the small angle approximation, we can derive a relationship between the collision angle and the impact parameter,

$$\Theta(b, \varepsilon_R) = \frac{b}{\varepsilon_R} \int_b^{\infty} \frac{1}{\sqrt{r^2 - b^2}} \left| \frac{dU}{dr} \right| dr. \tag{3.9}$$

If we use our generic form for the potential in equation (3.7), then $dU/dr = (-kC)/(r^{k-1})$. Plugging this into the integral, there is a general analytical solution of the form

$$\Theta(b, \varepsilon_R) = \frac{A(k)}{\varepsilon_R b^k}, \tag{3.10}$$

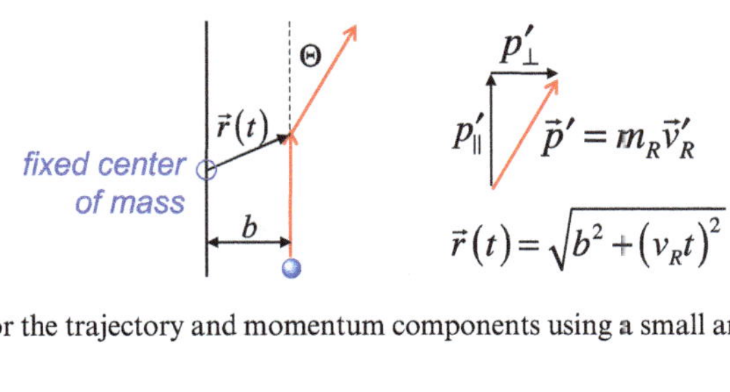

Figure 3.4. Schematic for the trajectory and momentum components using a small angle, straight line analysis.

[6] A dubious assumption but useful for demonstrating the relationship analytically.

where the function $A(k)$ is $A(k) = \frac{C\sqrt{\pi}\Gamma[(k+1)/2]}{2\Gamma[(k+2)/2]}$, with $\Gamma(x)$ being the Gamma function.

We now have a relationship between the impact parameter, scattering angle, and energy of the particles in the center-of-mass frame of reference. Applying equation (3.3) we can show that the differential scattering angle is

$$I(\varepsilon_R, \Theta) = \frac{1}{k}\left(\frac{A(k)}{\varepsilon_R}\right)^{2/k} \frac{1}{\Theta^{2+2/k}}. \tag{3.11}$$

Because the particles must scatter through the same effective solid angle regardless of the frame of reference, we can show that

$$2\pi I(\varepsilon_R, \Theta)\sin\Theta d\Theta = 2\pi I(\varepsilon_R, \theta_2)\sin\theta_2 d\theta_2. \tag{3.12}$$

As we can use basic frame-of-reference geometry to show that $\Theta = f(\theta_2)$, we can thus find $\frac{d\Theta}{d\theta_2}$. Therefore, to find the scattering cross-section as a function of energy, $\sigma(\varepsilon)$, we only need to find the scattering angle in the center-of-mass frame of reference.

The key takeaway is we now have an *explicit* form for the differential collision cross-section[7] that is a function of the relative energy of the two particles involved in the collision. What is crucial here is that since I is a function of ε_R, then σ is also a function of ε_R due to relationship in equation (3.4). The origin of this energy dependence is due to the interaction force between the particles from the interaction potential energy $U(r)$. Because the interaction depends on the local proximity between the two particles, it will also depend on the rate at which they approach each other, and thus their kinetic energy. It is important to note that this model is extremely simple and based on classical collision dynamics, whereas real gas particles and electrons will interact quantum mechanically. Still, the essential aspect, that real particle interactions lead to collision cross-sections being a function of kinetic energy, still holds.

Interestingly, with this approximation for $I(\varepsilon_R, \Theta)$, it is possible to derive approximations for the various types of electron–neutral collisions. One of the most famous is an approximation for the ionization collision cross-section based on the assumption that a free, incident electron has a binary collision with an electron in the outer shell of the target gas particle. This leads to the ionization cross-section as a function of the electron energy of [7, 11]

$$\sigma_{\text{ion}}(\varepsilon_e) = \pi\left(\frac{q^2}{4\pi\varepsilon_0}\right)^2 \frac{1}{\varepsilon_e}\left(\frac{1}{E_{\text{ion}}} - \frac{1}{\varepsilon_e}\right). \tag{3.13}$$

This classical treatment of ionization based on Newtonian mechanics is called the *Thomson cross-section*[8]. Figure 3.5 shows a plot of this cross-section for argon

[7] Under the assumptions of small angle scattering and a potential energy that decays with distance.

[8] Named after British physicist Sir Joseph John 'J J' Thomson, who was awarded the Nobel Prize in 1906 for the discovery of the electron and is also the inventor of the mass spectrometer.

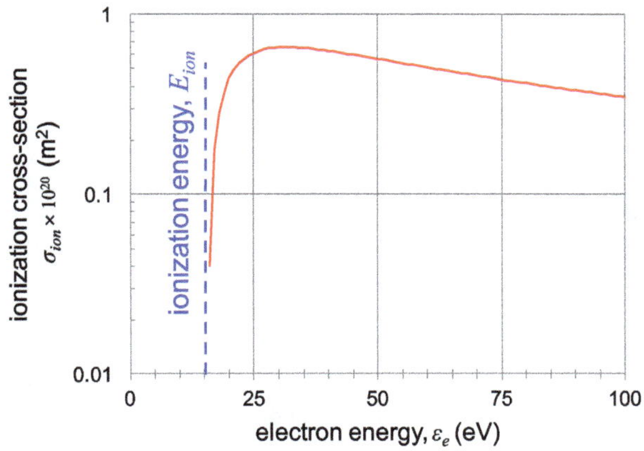

Figure 3.5. Thomson ionization cross-section for argon ($E_{\text{ion}} = 15.75$ eV).

($E_{\text{ion}} = 15.75$ eV). Clearly there is a maximum for this cross-section and we can calculate it to be

$$\sigma_{\text{ion, max}}(\varepsilon_e) = \frac{\pi}{4}\left(\frac{q^2}{4\pi\varepsilon_0}\right)^2 \frac{1}{E_{\text{ion}}^2}, \tag{3.14}$$

where it occurs at an electron energy of

$$\varepsilon_{e,\, \sigma_{\text{ion, max}}} = 2E_{\text{ion}}. \tag{3.15}$$

3.2.4 Realistic collision cross-sections

While these analytical approximations are useful for understanding why the collision cross-section is a function of the electron energy, they are not very accurate. To acquire accurate collision cross-sections either requires detailed measurements or more complex quantum models. However, for many different gases, electron collision cross-sections have been catalogued and published[9]. Consider, for example, figure 3.6, which shows the elastic, first electronic excitation, and ionization cross-section for argon. There are a few notable features. First, as expected, the elastic collision cross-section is non-zero for all electron energies, whereas the excitation and ionization cross-sections are only non-zero above their respective threshold energies. Additionally, the elastic collision cross-section is greater than its excitation or ionization counterparts for nearly all energies. As collision cross-section relates directly to reaction probability, it is clear the elastic collisions are far more common than inelastic collisions except for electrons that have relatively high energies

[9] See for example [5, 6]. Additionally, a database is now curated called the Plasma Data Exchange Project housed at https://fr.lxcat.net/ [9].

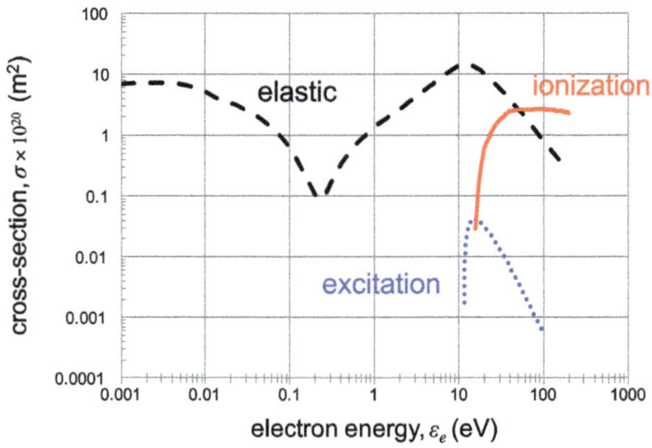

Figure 3.6. Electron–argon collision cross-sections as a function of electron energy from [10].

(greater than 70 eV or so). Finally, if we look specifically at the ionization cross-section, it peaks around 100 eV[10]. This is appreciably more than the peak predicted by the Thomson approximation in equation (3.15) ($2E_{\text{ion}} \approx 31.5$ eV). This highlights how simple analytical expressions fail to capture the true complexity of electron–neutral collisions; although they can produce the same shape of the cross-section curve, they cannot predict accurate values.

3.3 Electron energy distribution

It should be clear at this point that there are many electron–neutral collisions that can occur and that the probability of each one of these strongly depends on the energy of the incident electron. However, as we learned in chapter 2, we cannot just look at these collisions at the individual level, but also need to consider the collection of electrons in a discharge. And what these collisions serve to do is change the energies of the electrons. So while the electrons *gain* energy from an electric field, they *lose* energy due to inelastic collisions. In some cases it is only a little, such as when an electron vibrationally excites a neutral gas particle, and in some cases it is a lot, such as in an ionization collision. But, this redistribution of energies also affects the probabilities of subsequent collisions, which are energy-dependent. Hence, it is therefore imperative to understand the energy *distribution* of the electrons, analogous to how we defined the equilibrium energy distribution for gas particles in section 2.3 as the Maxwell–Boltzmann energy distribution. Unlike the particles in an equilibrium gas, however, the electrons in a low-temperature, non-equilibrium gas discharge are *not* in equilibrium with the gas particles (electrons give energy to the

[10] Interestingly, this is true for nearly every element—the peak collision cross-section occurs around 70–100 eV. For this reason, electron beam ionization sources such as in mass spectrometers almost always accelerate the electron beam to 70 eV to ensure the highest probability of ionization.

particles but not vice versa), and therefore the Maxwell–Boltzmann distribution does not strictly apply.

When considering the energy distribution of the electrons, we define a probability distribution function (PDF) $f_e(\varepsilon_e)$ that an electron will have a given energy, often called the electron energy probability function (EEPF), that follows

$$\int_0^\infty f_e(\varepsilon_e) d\varepsilon_e = 1. \tag{3.16}$$

Often, however, an alternative form, called the electron energy distribution function (EEDF) $f_{\text{EEDF}}(\varepsilon_e)$, is used, and defined as

$$f_{\text{EEDF}}(\varepsilon_e) = f_e(\varepsilon_e) \cdot \varepsilon_e^{-1/2}. \tag{3.17}$$

It is notable that the EEDF is not strictly a probability distribution function, which by definition is normalized to a value of 1 when integrated, but rather the normalization follows

$$\int_0^\infty f_{\text{EEDF}}(\varepsilon_e) \varepsilon_e^{1/2} d\varepsilon_e = 1. \tag{3.18}$$

Properly determining the EEPF or the EEDF in a low-temperature, non-equilibrium discharge can be exceedingly difficult as they are governed by the *seven-dimensional, integro-differential Boltzmann transport equation* (BTE), as is discussed in chapter 7. However, there are a few analytical forms that can be derived under simplified assumptions. The most common one is the *Maxwellian* distribution, which assumes a *constant collision frequency*[11] and takes the form

$$f_e(\varepsilon_e) = \frac{2}{\sqrt{\pi}} \left(\frac{1}{(k_B T_e)^3} \right)^{1/2} \varepsilon_e^{1/2} \exp\left(-\frac{\varepsilon_e}{k_B T_e} \right). \tag{3.19}$$

In the Maxwellian distribution, the energy is normalized by the parameter $k_B T_e$, where T_e is the electron temperature. As was true in equation (2.24), the average electron energy corresponds to

$$\overline{\varepsilon_e} = \frac{3}{2} k_B T_e, \tag{3.20}$$

and thus the electron temperature is a measure of the *average* electron energy. As written in equations (3.19) and (3.20), T_e has units of temperature (i.e. K). However, typically, T_e is reported in energy units of eV, which implies k_B is effectively 1 eV K^{-1} [4].

As you may notice, the Maxwellian is the same as the equilibrium Maxwell–Boltzmann distribution for gas particles in equation (2.23), and this is because it

[11] The collision frequency relates to the average time between collisions, as discussed in section 3.4.

assumes the electrons are in thermodynamic equilibrium with each other[12]. However, equilibrium requires a significant number of electron–electron collisions so that the electrons distribute their energies amongst each other, and this only occurs when there is a very high degree of ionization. In non-equilibrium, low-temperature discharges, this is not the case, and the distribution is strictly non-Maxwellian. In fact, the Maxwell–Boltzmann distribution assumes elastic collisions only, whereas inelastic electron–neutral collisions, which are 'heavy' particles to the electrons, sap the electrons of their energy, moving the EEPF away from this Maxwellian ideal. For this reason, there is a greater number of higher energy electrons in non-equilibrium, low-temperature discharges than what is described by the Maxwell–Boltzmann distribution [4]. This is often approximated by what is known as a two-temperature Maxwellian or bi-Maxwellian distribution, which can be written as simply two different Maxwellian distributions

$$f_e(\varepsilon_e) = \begin{cases} A_0 \varepsilon_e^{1/2} \exp\left(-\dfrac{\varepsilon_e}{k_B T_{e,\,l}}\right), & \varepsilon_e < \varepsilon_{\mathrm{ex},1} \\ A_0 \varepsilon_e^{1/2} \exp\left[-\varepsilon_{\mathrm{ex},\,1}\left(\dfrac{1}{k_B T_{e,\,l}} - \dfrac{1}{k_B T_{e,\,h}}\right)\right] \exp\left(-\dfrac{\varepsilon_e}{k_B T_{e,\,h}}\right), & \varepsilon_e > \varepsilon_{\mathrm{ex},1} \end{cases}. \quad (3.21)$$

The subscripts l and h refer to the average electron temperatures of the low- and high-energy electron populations, and the division between the two populations is typically assumed to be the first electronic excitation energy $\varepsilon_{\mathrm{ex},1}$. The parameter A_0 is a normalization parameter [4].

Figure 3.7 shows a comparison of Maxwellian and bi-Maxwellian distributions, and it is clear that they have distinct shapes. As shown in figure 3.6, ionization is *most* probable for electron energies ~70–100 eV. It is clear, therefore, that the

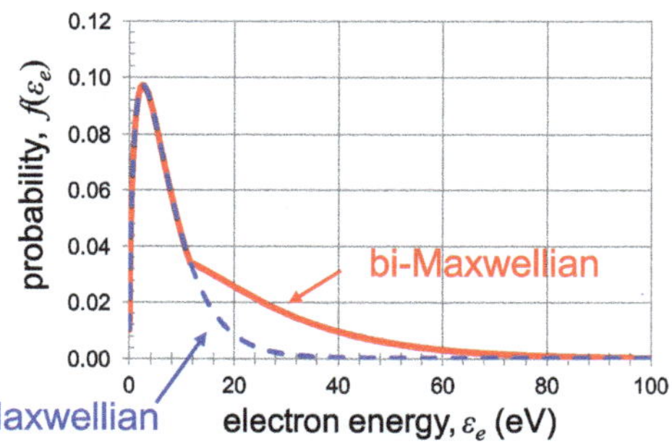

Figure 3.7. Maxwellian and two-temperature bi-Maxwellian distributions for argon assuming $T_e = T_{e,\,l} = 5$ eV and $T_{e,\,h} = 15$ eV.

[12] In fact, the quantity $\varepsilon_e^{1/2}$ in the pre-factor to the exponential is key here. By dividing it out, one moves from the EEPF to the EEDF such that $f_{\mathrm{EEDF}}(\varepsilon_e) \propto \exp(-\varepsilon_e/k_B T_e)$. Using this form is often easier, because one can simply take the logarithm of both sides to obtain a linear relationship.

high-energy tail described by the bi-Maxwellian distribution is key; it is those high-energy electrons that do most of the useful work in the discharge in terms of both ionization and excitation.

In addition to the Maxwellian and two-temperature Maxwellian, there are other analytical forms that have been derived, but just as with the above, are only approximate in nature. For example, the Druyvesteyn distribution [3] assumes a constant mean free path, rather than frequency, and this implies an energy-independent collision cross-section. Clearly this is not the case, as we have just proven in section 3.2. Similarly, the Margenau distribution [8] assumes a constant mean free path, but allows for an alternating field and thus is typically accurate for high-frequency discharges. Again, two-temperature approximations of these can also be applied [2]. But as mentioned before, the 'real' distribution is actually governed by the Boltzmann transport equation and rarely, if ever, follows any of these simplified analytical expressions exactly.

3.4 Electron collision frequency and reaction rates

As alluded to earlier, electron–neutral collisions really reflect electron–neutral chemistry, where the excitation and ionization of the neutral are essentially chemical reactions. We can describe this chemistry using a kinetic framework based on the collision cross-section, which will play an important role in understanding ionization.

The collision frequency ν, or the number of collisions undergone by an electron per unit time[13], depends on the probability that the electron hits a neutral molecule, which we already know depends on the collision cross-section $\sigma(\varepsilon_e)$, and the number of electrons that have that kinetic energy ε_e, which is described by $f_e(\varepsilon_e)$. We can treat this by an overlap integral of $\sigma(\varepsilon_e)$ and $f_e(\varepsilon_e)$ as

$$\nu = N_g \frac{\int_0^\infty f_e(\varepsilon_e)\sigma(\varepsilon_e)v_e d\varepsilon_e}{\int_0^\infty f_e(\varepsilon_e)d\varepsilon_e}, \tag{3.22}$$

where N_g is the number density of gas molecules[14] and v_e is the velocity of the impacting electron[15].

The ratio of only the integrals (that is, N_g is not considered) is, by definition, the energy average of the product of $\sigma(\varepsilon_e)$ and v_e or

$$\langle \sigma(\varepsilon_e)v_e \rangle = \frac{\int_0^\infty f_e(\varepsilon_e)\sigma(\varepsilon_e)v_e d\varepsilon_e}{\int_0^\infty f_e(\varepsilon_e)d\varepsilon_e}. \tag{3.23}$$

---

[13] Clearly, based on this definition, frequency has units of s^{-1}.

[14] The 'target' particles in these collisions.

[15] If we properly normalize our EEPF, $f_e(\varepsilon_e)$, the quantity in the denominator is $\int f_e(\varepsilon_e)d\varepsilon_e = 1$, as shown in equation (3.16). Frequently, the collision frequency and reaction rate coefficient are written as only the numerator.

Therefore, the collision frequency can also be written as

$$\nu = N_g\langle\sigma(\varepsilon_e)v_e\rangle. \tag{3.24}$$

We define the quantity as $\langle\sigma(\varepsilon_e)v_e\rangle$ as the reaction rate constant[16] k or

$$k = \langle\sigma(\varepsilon_e)v_e\rangle. \tag{3.25}$$

As written, this collision frequency considers *all* electron collisions, be it elastic, excitation, or ionization, and uses the **total collision cross-section**. The total collision cross-section is defined as the sum of all the cross-sections for each individual reaction,

$$\sigma(\varepsilon_e) = \sigma_{\text{el}}(\varepsilon_e) + \sum_j \sigma_{\text{ex},\,j}(\varepsilon_e) + \sum_k \sigma_{\text{ion},\,k}(\varepsilon_e). \tag{3.26}$$

Here, σ_{el} is the elastic collision cross-section, $\sigma_{\text{ex},\,j}$ the excitation cross-sections for each excitation reaction j, and $\sigma_{\text{ion},\,k}$ the ionization cross-sections for each ionization reaction k.[17] However, we can write the same expression for any specific *single* electron–neutral collision by simply considering only the collision cross-section for that reaction. That is, for reaction j, where j is a specific inelastic collision, we use the cross-section $\sigma_j(\varepsilon_e)$ such that the reaction frequency is

$$\nu_j = N_g \frac{\int_0^\infty f_e(\varepsilon_e)\sigma_j(\varepsilon_e)v_e d\varepsilon_e}{\int_0^\infty f_e(\varepsilon_e)d\varepsilon_e}. \tag{3.27}$$

Therefore, the reaction frequency can also be written as

$$\nu_j = N_g\langle\sigma_j(\varepsilon_e)v_e\rangle = N_g k_j, \tag{3.28}$$

where $k_j = \langle\sigma_j(\varepsilon_e)v_e\rangle$ defines the reaction rate constant for reaction j. Reaction rate constants can be defined for every type of excitation reaction as well as ionization. As we will learn in part II, these rate constants k can be plugged directly into rate balance equations that govern discharges.

It is notable that the collision frequency and collision rates can be related to the mean free path. One way, for example, of defining the mean free path is

$$\bar{l} = \frac{\bar{v}}{\nu}, \tag{3.29}$$

where $\bar{v}$ is the average speed of the particle. When we define the mean free path based on the cross-section, as in equation (2.13), we are using the energy-averaged, total collision cross-section ($\bar{\sigma}$). But we can similarly define elastic, excitation, and ionization mean free paths by replacing the σ by $\bar{\sigma}_{\text{el}}$, $\bar{\sigma}_{\text{ex}}$, or $\bar{\sigma}_{\text{ion}}$, respectively.

---

[16] The units of k are $m^3\ s^{-1}$.

[17] Note that there can be multiple types of ionization reactions depending on the gas type. For example, nitrogen gas N_2 has reactions that form both N_2^+ and N_2^{2+}.

In their current form, these collision frequency and reaction rate expressions are functions of ε_e and v_e. But we can convert them into *only* functions of ε_e by using the classic kinetic energy expression $\varepsilon = \frac{1}{2}mv^2$. If the energy is given in electronvolts, the reaction rate constant for reaction j can be defined as

$$k_j = \sqrt{\frac{2q}{m_e}} \int_0^\infty f_e(\varepsilon_e)\sigma_j(\varepsilon_e)\varepsilon_e^{1/2} d\varepsilon_e, \tag{3.30}$$

in terms of the EEPF or

$$k_j = \sqrt{\frac{2q}{m_e}} \int_0^\infty f_{\mathrm{EEDF}}(\varepsilon_e)\sigma_j(\varepsilon_e)\varepsilon_e d\varepsilon_e, \tag{3.31}$$

in terms of the EEDF.

3.5 Summary

The purpose of this chapter was to give an overview of the basic behavior of free electrons, starting with their collision dynamics and concluding with how this affects their energy distribution and ultimately electron-driven reactions. Some of the key points are as follows:

- Free gaseous electrons can collide with neutral gas atoms and molecules. These collisions can be elastic or inelastic, where inelastic collisions excite the atom or molecule into a higher rotational, vibrational, or electronic energy state or ionize the atom or molecule. These collisions correspond to electron–neutral chemical reactions.
- The collision cross-section for electron–neutral collisions represents the probability of a free electron colliding with a gas particle. They depend on the kinetic energy of the incident electron, and each collision type—elastic or inelastic—has its own energy-dependent collision cross-section. The energy dependence arises from the detailed interaction of the incident electron with the nucleus and bound electrons of the gas particle and the force between these entities.
- Electron–neutral collisions redistribute the energy of free gaseous electrons, as some of their kinetic energy is given to the gas particles via inelastic collisions. The energy distribution of the collection of free electrons is called the electron energy probability function (EEPF) or electron energy distribution function (EEDF), and is often non-Maxwellian in non-equilibrium, low-temperature discharges.
- The frequency of a specific electron–neutral collision, as well as the corresponding reaction rate constant, is based on the energy-dependent collision cross-section and energy distribution of the electron population. The total collision frequency is based on the total collision cross-section, which is the sum (at every energy) of each individual collision.

Electrons are, by and large, *the* most important particle in non-equilibrium, low-temperature discharges, and electron–neutral collisions are the source for nearly all the important physics and chemistry. For this reason, understanding (and predicting) the behavior of electrons is critical in the design and operation of gas discharges for nearly every application.

References

[1] Anderson R V 1966 Absolute measurements of atmospheric charge density *J. Geophys. Res.* **71** 5809–14

[2] Carman R J and Mildren R P 2000 Electron energy distribution functions for modelling the plasma kinetics in dielectric barrier discharges *J. Phys. D: Appl. Phys.* **33** L99–103

[3] Druyvesteyn M J 1930 Der Niedervoltbogen (The low-voltage arc) *Z. Phys.* **64** 781–98

[4] Fridman A and Kennedy L A 2011 *Plasma Physics and Engineering* 2nd edn (Boca Raton, FL: CRC Press)

[5] Itikawa Y, Hayashi M, Ichimura A, Onda K, Sakimoto K, Takayanagi K, Nakamura M, Nishimura H and Takayanagi T 1986 Cross sections for collisions of electrons and photons with nitrogen molecules *J. Phys. Chem. Ref. Data* **15** 985–1010

[6] Itikawa Y, Ichimura A, Onda K, Sakimoto K, Takayanagi K, Hatano Y, Hayashi M, Nishimura H and Tsurubuchi S 1989 Cross sections for collisions of electrons and photons with oxygen molecules *J. Phys. Chem. Ref. Data* **18** 23–42

[7] Lieberman M A and Lichtenberg A J 2005 *Principles of Plasma Discharges and Materials Processing* 2nd edn (New York: Wiley)

[8] Margenau H 1946 Conduction and dispersion of ionized gases at high frequencies *Phys. Rev.* **69** 508–13

[9] Pitchford L C 2017 LXCat: an open-access, web-based platform for data needed for modeling low temperature plasmas *Plasma Process Polym.* **14** 1600098

[10] Puech V and Torchin L 1986 Collision cross sections and electron swarm parameters in argon *J. Phys. D: Appl. Phys.* **19** 2309–23

[11] Thomson J J 1912 Ionization by moving electrified particles *Philos. Mag.* **23** 449–57

IOP Publishing

Ionization and Ion Transport (Second Edition)
A primer for the study of gas discharges and plasmas
David B Go

Chapter 4

Gaseous ionization processes

With a firm foundation in basic concepts from gas dynamics and a more focused understanding of basic electron behaviour, we can now begin a more fruitful inspection of ionization. Recall that here, we are referring to ionization in the broad sense of creating the charges—both electrons and ions—necessary to convert an insulating gas into a conductive discharge. We can broadly classify these into two types of processes—those that occur *in the gas* and those that occur *at a surface*. In the former, the atoms and molecules are made to be charged, producing gaseous free electrons and ions. In the latter, charges, and specifically electrons, are *injected* into the gas by a surface, most often an electrode. In both cases, the result is freely moving charges in the gas phase, enabling conduction, but the origin of these charges is substantially different. In this chapter, we focus specifically on the various gaseous ionization processes that may occur.

4.1 Electron-impact ionization

As introduced in chapter 3, electrons are the most important particle in a gas discharge, and they play a particularly important role in ionization. Suppose that we have a single free electron in a gas and apply an electric field. The electron will be accelerated according to equation (1.2), and if it obtains a kinetic energy ε_e greater than the ionization energy E_{ion} of a gas particle, there is some probability that it will ionize the gas particle, producing a second free electron. That is, through a single **electron-impact ionization** event, we have already doubled the number of free electrons in the gas.

As discussed in section 3.4 we can describe the frequency of any specific type of collision based on knowledge of the collision cross-section and electron energy distribution. Applying equation (3.27) and only considering the ionization cross-section $\sigma_{\text{ion}}(\varepsilon_e)$, the *ionization frequency* due to electron-impact is

$$\nu_{\text{ion}} = N_g \frac{\int_0^\infty f(\varepsilon_e)\sigma_{\text{ion}}(\varepsilon_e)v_e d\varepsilon_e}{\int_0^\infty f(\varepsilon_e)d\varepsilon_e}. \tag{4.1}$$

doi:10.1088/978-0-7503-3991-9ch4

The ionization frequency can also be written as

$$\nu_{\text{ion}} = N_g \langle \sigma_{\text{ion}}(\varepsilon_e) v_e \rangle = N_g k_{\text{ion}}, \tag{4.2}$$

where $k_{\text{ion}} = \langle \sigma_{\text{ion}}(\varepsilon_e) v_e \rangle$ defines the ionization reaction rate constant. Simply put, ν_{ion} is the number of ionization events performed by a single free electron per second, or the rate of ion production.

4.1.1 Temporal electron avalanche

With the ionization frequency now defined, the overall (or net) creation of electrons (recall each ionization event also creates an electron) can now be considered. The rate of creation of electrons by free electrons is simply the ionization rate for a single free electron ν_{ion} multiplied by the number of free electrons N_e.[1] Mathematically, this rate can be written as the time rate-of-change of the number of free electrons or

$$\frac{dN_e}{dt} = \nu_{\text{ion}} N_e. \tag{4.3}$$

If it is assumed that ν_{ion} is constant, then this first-order differential equation is easily solved,

$$N_e(t) = N_{e,0} e^{\nu_{\text{ion}} t}, \tag{4.4}$$

where $N_{e,0} = N_e(t = 0)$, or the number of free electrons at time zero. Equation (4.4) implies that there is an *exponential* growth in the number of electrons over time—this is called an **electron avalanche**. However, it is not necessary to only think of the avalanche in time. In fact, it is often much easier to picture it in space.

4.1.2 Spatial electron avalanche

If the number of electrons created per second is defined as ν_{ion}, a similar concept can be defined in space. That is, the number of electrons created by a single electron as it travels a specified distance (e.g. one centimeter); we define a new parameter for this that we call α.[2] The ratio of these two, ν_{ion}/α, simply has the units of velocity. This is called the *drift* or mean velocity of an electron v_e, and we can write this formally as

$$\nu_{\text{ion}} = \alpha v_e. \tag{4.5}$$

Note that ν_{ion} is the 'true' description of the ionization rate—it is *defined* by the collision probability and energy distribution. In contrast, α is a derived quantity literally defined by equation (4.5). However, α is fairly easy to measure, and it is in

[1] Technically, the number density of free electrons as the units of N_e are m^{-3}.

[2] Thus, the units of α are cm^{-1}. The use of centimeters as the unit of distance here is arbitrary and simply the standard convention. α could just as easily be defined over a meter, mile, or light-year.

many ways a more convenient way to think about electron-impact ionization[3]. The parameter α is called **Townsend's first ionization coefficient**[4]. Similar to how a temporal electron avalanche was defined based on ν_{ion}, a spatial electron avalanche can be described using α.

Consider the general schematic of a dc gas discharge shown in figure 4.1. There are two electrodes connected by a dc power source, the *negatively biased* **cathode** and the *positively biased* **anode**. Due to the applied voltage V_A from the power supply, the cathode is at a negative voltage relative to the anode, and there is a potential energy drop ($\Delta\Phi = V_A$) between the electrodes. As described in chapter 1, this voltage difference produces an electric field where, in one dimension (1D), the electric field is defined as

$$E_x = -\frac{d\Phi}{dx}. \tag{4.6}$$

If there are few ions or free electrons between the electrodes, this electric field is simply the applied voltage V_A divided by the distance between the electrodes d, or $E_x \approx -V_A/d$. This electric field thus points from the anode toward the cathode and any positive ions between the electrodes will be accelerated toward the cathode. Conversely, any electrons or negative ions in the domain will move toward the anode.

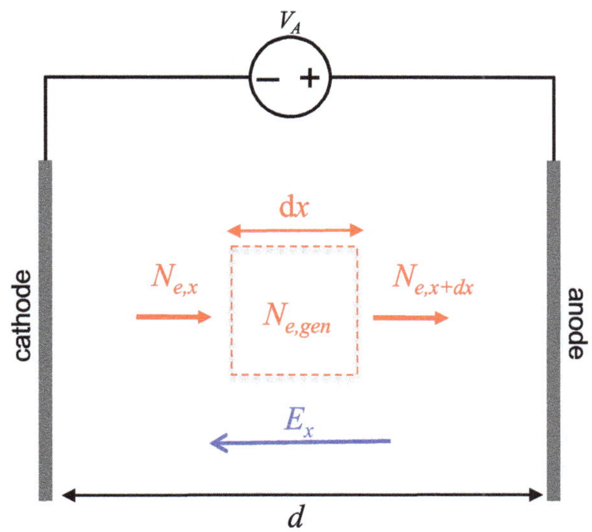

Figure 4.1. Schematic of a dc circuit between a cathode (negative) and anode (positive) and a control volume for the electron transport.

[3] For this reason, electron-impact ionization is often referred to as the α-process.

[4] Named after pioneering plasma scientist, the Irishman Sir John Sealy Townsend, who essentially discovered the electron avalanche and wrote one of the first books on the topic [1].

Now consider a small, 1D control volume of size dx between the electrodes. Because electrons are moving toward the anode, some number of electrons $N_{e,x}$ will enter the control volume at x, and some new number $N_{e,x+dx}$ will leave the control volume at $x + dx$. Because of conservation, the difference between these two must be equal to the number created (or destroyed) within the control volume, $N_{e,\text{gen}}$. Over the distance dx, a single electron creates α new electrons. Therefore, the total number generated by $N_{e,x}$ entering the control volume is $N_{e,\text{gen}} = N_{e,x}\alpha dx$. Based on the balance of electrons entering and exiting the control volume with those generated by electron-impact ionization, we can derive a differential equation for the evolution of the electrons in space of the form

$$\frac{dN_e}{dx} = \alpha N_e. \tag{4.7}$$

It should be readily apparent that equation (4.7) is the spatial analog to equation (4.3). If some number of electrons $N_{e,0}$ are initiated at the cathode ($x = 0$), the solution to this equation is

$$N_e(x) = N_{e,0}e^{\alpha x}. \tag{4.8}$$

This is also called an **electron avalanche**—but now it is in space. That is, a single electron produces exponentially more electrons as it traverses a gap from the cathode to the anode, as shown in figure 4.2. In fact, if we integrate equation (4.7) from the cathode to the anode (i.e. from $x = 0$ to $x = d$), we obtain the total number of electrons produced in the electrode gap electrons initiated at the cathode,

$$N_{e,d} = N_{e,0}\, e^{\alpha d}. \tag{4.9}$$

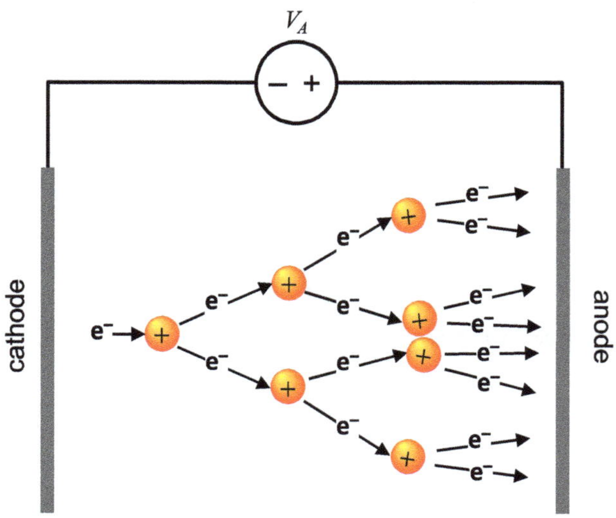

Figure 4.2. Sketch of an electron avalanche.

We can relate this equation for the number density N_e to the current by recognizing that the **current density** can be defined as $j_e = qN_e v_e$, where q is the charge of the electrons and as noted above, v_e is drift velocity[5]. The electron **current** i is equivalently the current density multiplied by the surface area of the electrodes (cathode or anode) and thus

$$i = j_e A = qN_e v_e A_{\text{electrode}}. \tag{4.10}$$

With this relationship (as long as the velocity is the same everywhere, which is a reasonable first approximation), we can rewrite equation (4.8) in terms of electron current. Thus, the electron current passing through the anode is simply

$$i_{\text{anode}} = i_{e,0} e^{\alpha d}, \tag{4.11}$$

where $i_{e,0}$ is the electron current leaving the cathode (due to an emission process, as discussed in chapter 5). By taking the logarithm of both sides, the current in equation (4.11) can be rewritten as

$$\ln(i_{\text{anode}}) = \alpha d + C, \tag{4.12}$$

where C is some constant.

In practice, α is determined experimentally by varying d but maintaining a constant E (that is, the applied voltage V_A is also varied based on equation (4.6)). The measured current at the anode can then by plotted as $\ln(i_{\text{anode}})$ versus d, and extracting the slope gives a value for α.[6]

4.1.3 Townsend's first ionization coefficient α

As stated previously, α is called **Townsend's first ionization coefficient** and electron-impact ionization is often referred to as the α-process. Physically, α is *the number of ions created by a single electron over some distance.* Further, it should be clear that the number of ionization collisions will be related to how fast the electrons are accelerated because the collision cross-section is energy-dependent (as discussed in section 3.1). Additionally, it makes conceptual sense that the number of ions generated should depend on how many gas particles are available for ionization (i.e. the gas density). Therefore, we can see that α should be a function of the electric field E, which provides the force that accelerates the electrons to suitable kinetic energies, and the gas pressure p, which defines the number of gas atoms. In fact, what is critical is the ratio of the electric field to the pressure, E/p, which is often called the **reduced electric field**. Ultimately, a *similarity law* describes α to be p multiplied by some function of E/p,

$$\alpha = p{\cdot}f(E/p). \tag{4.13}$$

---

[5] Recall that the current density takes units of A m^{-2}.

[6] Some good papers that show how α was measured in the early years of gas discharge study were written by Frederick Sanders. See, for example, [2, 3].

Therefore, an alternative approach to experimentally determining α is to vary p and maintain E/p as a constant, and then analyze the resulting current data using the relation

$$\ln(i_{\text{anode}}) = \frac{\alpha}{p}(pd) + C. \tag{4.14}$$

It should be noted that the product pressure times distance pd and the reduced pressure E/p are both crucial scaling factors in gas discharges and form the basis for important scaling relationships[7].

The most common form of α, which has been empirically validated for specific conditions, takes the form[8]

$$\alpha = Ap \exp\left[-\frac{Bp}{E}\right], \tag{4.15}$$

where A and B are constants that depend on the gas composition[9]. Other empirical forms include

$$\alpha = Cp\exp\left[-D\sqrt{\frac{p}{E}}\right] \tag{4.16}$$

for noble gases and

$$\alpha = 1.17 \times 10^{-4} p\left(\frac{E}{p} - 32.2\right)^2 \tag{4.17}$$

for air in the reduced electric field range of $E\ /\ p = 44$–176 V cm^{-1} Torr^{-1}.

4.2 Ion-impact ionization

Electron-impact ionization (the α-process), where an electron collides with a neutral particle to create a second free electron and positive ion, is the dominant gaseous ionization mechanism. However, it is reasonable to anticipate that positive ions accelerated by the electric field would *also* collide with neutral molecules to form a second ion and a free electron. This is called the β-process where β is Townsend's second ionization coefficient and is defined in the same manner as α (the number of ions a single ion creates over some distance). Using a similar control volume approach as shown in figure 4.1, a new differential equation can be obtained that describes both the α- and β-processes together to generate positive ions N_i,

$$\frac{dN_i}{dx} = \alpha(N_{e,0} + N_i) + \beta N_{i,0}. \tag{4.18}$$

[7] This will be discussed more in chapter 8.

[8] This relationship can be derived phenomenologically by considering the mean free path and accelerating electric field. The constants A and B are typically determined empirically, but can be approximated by $A = 1/(p\bar{l})$ and $B = A/E_{\text{ion}}$, where $\bar{l}$ is the mean free path and E_{ion} is the ionization energy [4].

[9] In common gas discharge parlance, A takes units of cm^{-1} Torr^{-1} and B takes units of V cm^{-1} Torr^{-1}. For a list of typical values, see table 4.1 of Raizer [5].

Here, $N_{e,0}$ is the initial number of electrons and $N_{i,0}$ is some initial number of positive ions (presumably created by earlier electron-impact collisions before entering the control volume). By considering that the total number of electrons N_e created across the electrode gap as simply the sum of the initial number $N_{e,0}$ and the total number of ions $N_{i,0} + N_i$,

$$N_e = N_{e,0} + N_{i,0} + N_i, \tag{4.19}$$

and plugging this into equation (4.18), an equation similar to equation (4.8) can be derived,

$$N_e(x) = N_{e,0}\frac{(\alpha - \beta)e^{(\alpha-\beta)x}}{\alpha - \beta e^{(\alpha-\beta)x}}. \tag{4.20}$$

This equation also shows an exponential avalanche production of electrons but considers both electron- and ion-impact events. Originally derived by Townsend [6], it was assumed that the β-process was equally as relevant as the α-process in gas discharges. However, this ultimately turned out to not be true [4]. For electron impact ionization, the kinetic energy of the electron must exceed the ionization energy E_{ion} of the neutral. However, for ions to ionize a neutral it must have a kinetic energy at least $2E_{\text{ion}}$. This is because the electric field of the ion itself is sufficiently strong to polarize the neutral causing a much 'softer' collision. In most gases where $E_{\text{ion}} \sim 10$ eV and ionization is most probable at $\varepsilon_e \sim 70$ eV, a positive ion requires approximately 400 eV in order to ionize a neutral. Further, ions move much slower than electrons (consider that the mass of a proton alone is three orders of magnitude greater than that of an electron). Therefore, for ions to readily contribute to ionization, they must be accelerated to much higher energies, and this takes much more time than accelerating their electron counterparts. The fact is that the time scale of discharge formation is much too fast for ions to contribute to the ionization process. Therefore, while ion-impact ionization is not non-existent, it plays a minor role in discharge formation—particularly in non-equilibrium, low-temperature discharges—and it is considered a negligible effect.

4.3 Photoionization

Similar to electron- and ion-impact, photons (from either external sources or from the plasma glow itself) of sufficient energy can ionize a neutral atom as well as excite it to a higher state or even dissociate a polyatomic molecule. The energy of a photon is

$$\varepsilon_{\text{photon}} = h\nu = \frac{hc}{\lambda}, \tag{4.21}$$

where h is Planck's constant, ν is the frequency of the light[10], λ is the wavelength of the light, and c is the speed of light[11]. The photo-reactions therefore are

[10] In this formulation, ν has units of hertz (Hz). If angular frequency ω rad s^{-1} is considered instead, then the reduced Planck's constant $\hbar = h/2\pi$ is used to obtain $\varepsilon_{\text{photon}} = \hbar\omega$.

[11] $c = 299\,792\,458$ m s^{-1}.

Excitation: $h\nu + M \rightarrow M^*$ iff $h\nu \geqslant E_{M^*}$.

Ionization: $h\nu + M \rightarrow M^+ + e^-$ iff $h\nu \geqslant E_{\text{ion}}$.

The question remains whether photoionization is relevant. Consider cesium vapor, which is the easiest element to ionize with an ionization threshold of only $E_{\text{ion}} = 3.87$ eV. Substituting into equation (4.21), the wavelength of light necessary to ionize cesium is $\lambda \approx 320$ nm, which is in the ultraviolet (UV) part of the electromagnetic spectrum. For any other atom or molecule, more energy is required (smaller wavelengths), so photoionization is, for the most part, an ultraviolet process.

One might conclude from this simple calculation that for all gases, ionization only occurs for $\lambda < 320$ nm. This is not quite right; it *is* possible to ionize gas at lower photon energies ($\lambda > 320$ nm), through a process called **step ionization**. Ionization can occur in *steps* where the neutral is first excited to a metastable state, and then a photon energizes the excited neutral to the ionization threshold, producing an ion. Therefore, the photon does not require the threshold energy E_{ion} because the neutral is already in an excited state. Therefore, it clear that while direct photoionization is unlikely, the step photoionization process can play an important role under the right circumstances.

So where do the photons come from? They can come from an external source (e.g. a laser) or from the gas itself. Recall, when an excited neutral relaxes back to its ground state, it emits a photon with $(h\nu)_{\text{emit}} = \varepsilon_{M^*}$. This is called resonance radiation. In non-equilibrium, low-temperature plasmas, photoionization is typically a secondary effect and less significant than electron-impact ionization, though not negligible. Corona discharges, in particular, are an example of a non-equilibrium, low-temperature discharge where photoionization is important, as photon-ionization produces seed electrons that then produce electron avalanches (and ultimately streamers) via typical electron-impact mechanisms.

4.4 Thermal ionization

This is also similar to electron-impact ionization but is simply neutral-impact ionization. The average kinetic energy of a gas particle can be approximated by $\varepsilon_M \sim k_B T$. From an order-of-magnitude analysis, the temperature of the gas needs to be, $T \sim 10^5$ K for a neutral to have kinetic energy that approaches that of the ionization energy $E_{\text{ion}} \sim 10$ eV. So it is clear that for this phenomena to be important the gas must be at extreme temperatures, and it is not relevant for non-equilibrium, low-temperature plasmas.

4.5 Step ionization

Step ionization, also called *stepwise ionization*, is when an electron first excites a gas particle to an excited state and then the excited particle is ionized by collision with a second electron. It is the electron-impact-only analog to photon-driven step ionization discussed above and can described by two reactions:

Excitation: $e^- + M \rightarrow M^*$ (step 1).

Ionization: $e^- + M^* \rightarrow 2e^- + M^+$ (step 2).

As written, this set of reactions shows a two-step ionization process. However, it can proceed along multiple steps, progressing through different excited states. As long as the electron and gas density are sufficiently high, stepwise ionization processes can be very important because there will be a correspondingly high concentration of excited gas neutrals. It can be shown that the ratio of reaction rate constants for stepwise ($k_{\text{ion,step}}$) and direct electron-impact ionization (k_{ion}) can be as high as $k_{\text{ion,step}}/k_{\text{ion}} \sim 10^3 - 10^4$ for equilibrium (thermal) discharges [7]. Non-equilibrium, low-temperature discharges have considerably lower electron concentrations, and excited neutrals will often probabilistically relax back to their ground state before experiencing a second electron collision. However, non-equilibrium, low-temperature discharges also have electron temperatures less than the ionization energy for most gases. Therefore, it is energetically easier for ionization to occur by step processes. Thus, this process, while not negligible, is not as dominant in non-equilibrium, low-temperature discharges as it is in equilibrium discharges and depends on the exact conditions and gases in the non-equilibrium, low-temperature discharge [8].

4.6 Penning ionization

Penning ionization[12] is a common gaseous ionization process that occurs in mixtures of two different gases called *Penning mixtures*. In this case, the ionization is not by ion or electron-impact, but by an excited neutral. An excited neutral of one species (X^*) collides with a second neutral of a different species (M), and during the energy exchange an electron is ejected from the target neutral:

$$X^* + M \rightarrow X + M^+ + e^-.$$

Penning ionization occurs when gas X has a higher ionization energy than gas M, that is, $E_{\text{ion},X} > E_{\text{ion},M}$, but gas X also has a metastable (long lasting) excited state X^* that has a greater energy than the ionization energy of the gas, or $E_{X^*} \geqslant E_{\text{ion},M}$.

For example, consider a mixture of argon and neon. The ionization energy of argon is $E_{\text{ion,Ar}} = 15.7$ eV whereas the ionization energy of neon is $E_{\text{ion,Ne}} = 21.6$ eV. However, neon has a metastable state that has an excitation energy of $E_{\text{Ne}^*} = 16.6$ eV. Therefore when a metastable Ne^* and neutral argon collide, there will be an energy exchange as the Ne^* relaxes to its ground state, and that energy is sufficient to ionize the argon atom. However, because of the complex thermodynamics of discharges where all the reaction processes affect energy distributions of the various particles (electrons, ions, etc), this type of Penning reactions is best only when the gas admixture is Ne + 0.1% Ar, as opposed to 10% Ar for example. In this case, the excited state of argon has an energy of $E_{\text{Ar}^*} = 11.5$ eV, such that increasing the argon concentration increases the number of Ar^*, but quenches the production of Ne^* thereby quenching Penning ionization.

Typical Penning gas admixtures include argon in neon, krypton in argon, krypton in neon, xenon in argon, and xenon in helium. The most prominent impact of using a

[12] Named after Dutch physicist Frans Michel Penning, who first reported it in 1927 [9].

Penning gas rather than a pure gas is that the overall current will be increased due to the increased ionization. Typically, it is most effective to be at low reduced electric fields ($E/p \sim 0.5$–30 V cm^{-1} Torr^{-1}) to obtain the greatest benefit from the Penning effect, and under these optimal conditions, discharges of Penning mixtures can be more energy-efficient than in pure gases [4].

4.7 Electron attachment

As discussed in section 3.1, some gases are electronegative and therefore prefer to have an additional electron (negative ion) rather than losing an electron, with the most common being oxygen (O_2) and sulfur hexafluoride (SF_6) as well as halogens such as chlorine (Cl_2) and fluorine (F_2) and halogen-containing chemicals. The general reaction is

$$M + e^- \rightarrow M^-,$$

but a dissociative reaction is also possible,

$$XM + e^- \rightarrow X + M^-.$$

Sometimes an electron attachment coefficient is defined; denoted as η, it is the analog to α. It is fairly straightforward to derive the electron avalanche including an electron attachment coefficient, but that is not included here. Clearly, electron attachment is an important process for gas discharges in air as well as discharge configurations in microelectronics fabrication as halogens are good for reactive ion etching.

4.8 Summary

The purpose of this chapter was to give an overview of the creation of electrons *in the gas phase* that help turn an insulating gas into a conductive gas. The following processes were discussed:

- The α-process, also called electron-impact ionization, is the dominant mode of ionization that leads to an electron avalanche (i.e. the exponential growth in the ion concentration). Townsend's ionization coefficient α depends not only on the gas type but also the electric field and pressure.
- The β-process, also called ion-impact ionization, and thermal ionization are largely negligible in non-equilibrium, low-temperature gas discharges.
- Photoionization occurs when an atom or molecule absorbs a photon with energy greater than the ionization threshold. It can occur either directly or through a step process, and can be important under the right circumstances (e.g. for seeding electron avalanches in corona discharges).
- Step and Penning ionization processes occur when there are intermediate excited states that help facilitate the ionization process. Step ionization occurs in a pure gas, whereas Penning ionization occurs in a gas admixture (Penning mixture), and, under the right conditions, these multistep ionization processes can be important.

- In electronegative gases, electron attachment or the η process can be especially important, and rather than forming positive ions, it results in negative ions.

It is important to remember that all of the relevant ionization processes depend on electron-driven reactions, either direct ionization or the production of excited states, and therefore all of these processes are probabilistic. For instance, although an electron with kinetic energy greater that the ionization energy may create an ion through the α process, it is not guaranteed that this type of collision will occur—an elastic or excitation reaction could occur instead. As we learned in chapter 3, energy-dependent collision cross-sections determine the probability of any given reaction and ultimately the reaction rate for all of the different reactions. It is often true that one or two of these processes will be dominant, but this depends on the exact conditions of the discharge—the pressure, applied potential and electric field, and the gas mixture. It is also important to note that the different processes discussed here are not all-encompassing. There are other processes that occur in nature, such as associative ionization, associative Penning ionization, and ionization by nuclear radiation. However, the ones discussed here are the most relevant for the majority of non-equilibrium, low-temperature gas discharges.

References

[1] Townsend J S 1910 *The Theory of Ionization of Gases by Collision* (London: Constable)

[2] Sanders F H 1932 The value of the Townsend coefficient for ionization by collision at large plate distances and near atmospheric pressure *Phys. Rev.* **41** 667–77

[3] Sanders F H 1933 Measurement of the Townsend coefficients for ionization by collision *Phys. Rev.* **44** 1020–4

[4] Nassar E 1971 *Fundamentals of Gaseous Ionization and Plasma Electronics* (New York: Wiley)

[5] Raizer Y P 1991 *Gas Discharge Physics* (Berlin: Springer)

[6] Townsend J S 1903 LXXVI. The genesis of ions by the motion of positive ions in a gas, and a theory of the sparking potential *Philos. Mag.* **6** 596–618

[7] Kunc J A 1984 Stepwise ionization in a non-equilibrium, steady-state hydrogen plasma *J. Quant. Spectrosc. Radiat. Transfer* **32** 311–23

[8] Fridman A and Kennedy L A 2011 *Plasma Physics and Engineering* 2nd edn (Boca Raton, FL: CRC Press)

[9] Penning F M 1927 About ionization by metastable atoms *Nat. Sci.* **15** 818

Chapter 5

Electrode processes

With an understanding of the ionization processes that occur in the gas phase, we now turn our attention to the processes that occur at the electrodes to produce charges in the discharge. When we consider electrode charge production, we are considering the production of free, gas-phase electrons *from a surface*. Under appropriate conditions, electrons can be ejected from the surface of a metal (or semiconductor) into the gas phase, and this is called electron emission. In the formation of a discharge, these electrons will subsequently participate in electron-impact ionization (α-process) and excitation reactions, producing additional free electrons through the wide variety of mechanisms discussed in chapter 4; all of these contribute to the overall electron current across the electrode gap. Electrode emission processes are very important in the formation of gas discharges, and this chapter focuses on the fundamental physics behind electron emission and the various emission processes important to gas discharges.

5.1 Importance of electron emission

Recall from chapter 4 that the derivation of the electron avalanche required some amount of seed electrons at the cathode in order to initiate electron-impact ionization events (see $N_{e,0}$ in equation (4.8), for example). These cathode electrons are not only needed to initiate electron-impact ionization, but are necessary in order to ensure that the current in the discharge does not terminate. Consider, for example, the basic configuration of a dc gas discharge shown in figure 4.1, and reproduced here in figure 5.1. Electrons and ions produced between the two electrodes will clearly move in opposite directions toward the anode and cathode, respectively. Once they reach their respective electrodes, they are neutralized and charge is lost from the gas. In order to *sustain* the discharge and maintain current it is necessary to *inject* electrons from the external surfaces into the domains; otherwise, all the electrons in the gas will simply be lost and current will terminate. This can occur at any wall in the system, in principle, but in a dc configuration such

doi:10.1088/978-0-7503-3991-9ch5

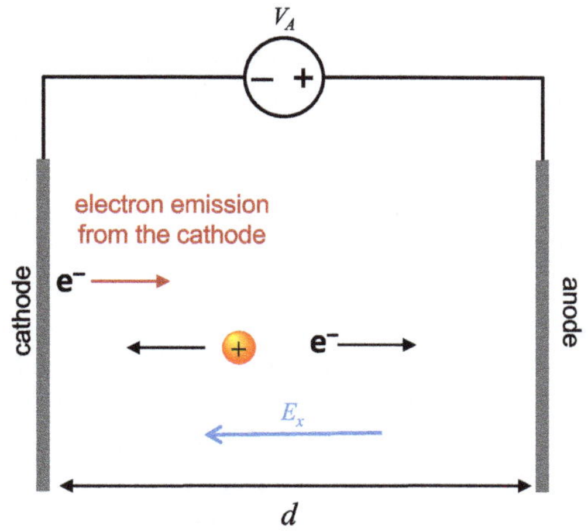

Figure 5.1. Schematic of a dc circuit between a cathode (negative) and anode (positive) showing the motion of electrons and ions and emission of electrons from the cathode (red).

as this, it will occur at the cathode. That is, electron emission from the cathode produces the gas-phase electrons necessary to sustain current and the discharge. Therefore, it is clear that electron emission is not only an important process but a *necessary* process for many types of non-equilibrium, low-temperature discharges to even work.

5.2 Foundational concepts

In order to understand electron emission, we should make sure we understand the basics of solid state physics that allows electrons to be ejected from a surface. This will then set the stage for understanding how and why electrons are emitted into gas discharges. Here we will discuss the important concepts for *metal* solids only, and leave detailed discussions into *insulating* and *semiconducting* materials for other texts[1].

5.2.1 The solid state

Generally, solid state physics is the study of solid matter, typically from an atomic perspective[2]. Conceptually, a crystalline solid[3] has its constituent atoms arranged in a *crystal lattice*, which is a periodic structure for the atoms in the solid that can be

[1] Two of the best texts on these foundational concepts for all three materials are *Solid State Physics* by Ashcroft and Mermin [1] and *Introduction to Solid State Physics* by Kittel [2].

[2] As opposed to solid mechanics, which studies solid matter from a continuum perspective.

[3] The other alternative is an amorphous solid. The primary difference is that a crystal has long-range order, whereas an amorphous solid has either no order or only short-range order. That is, the atoms in a crystal are arranged in a regular, repeated pattern, whereas the atoms in an amorphous solid have no discernible pattern. We will only consider crystals here.

visualized as a complex spring–mass system with the springs representing the bonds between neighboring atoms and the masses of the atoms themselves.

From the view of a single atom, each electron associated with the atom has a potential energy due to local attraction to the protons in the nucleus. As discussed in chapter 2, electrons inhabit discrete, quantized energy states called orbitals that can be described by four quantum numbers. In the atomic structure, the valence electrons are in the outermost shells such that they are loosely bound to the nucleus and therefore free to move. Most metals only have partially filled outer shells such that there is 'space' for these valence electrons to move. In a crystal structure where atoms are bonded together, these outer orbits are hybridized, and the atoms now effectively share these electrons. Thus when a force is applied to the electrons, such as an electric or magnetic field, the electrons are free to move from atom to atom, which is what makes metals electrically conductive[4].

For metals, the crystal lattice forms the backbone of the solid for these free-to-move outer valence electrons. We can conceptually visualize these electrons in the outer shell as free, independent particles, and treat them as though they are not influenced by the atoms in the crystal lattice, the so-called free electron model[5]. However, these electrons are still governed by quantum mechanics and have discrete available energy levels. Here an extremely simplified view of these concepts will be described in order to set the stage for understanding electron emission.

The energy and momentum of the free electrons are described by Schrödinger's equation,

$$\frac{-\hbar^2}{2m_e}\nabla^2\Psi(\vec{r}) = E\Psi(\vec{r}), \tag{5.1}$$

where $\hbar$ is the reduced Plank's constant, m_e is the electron mass, E is the energy, and Ψ is the *electron wavefuction.* The wavefunction $|\Psi^2|$ can be thought of as an electron probability or likelihood of an electron being at the position $\vec{r}$. A general solution to Schrödinger's equation is

$$\Psi(\vec{r}) = \frac{1}{\sqrt{V}}e^{i\vec{k}\cdot\vec{r}}, \tag{5.2}$$

where V is a volume and $\vec{k}$ is a wave vector. The energy of the electrons comes from the *eigenfunctions* of Schrödinger's equation and are a function of the wave vector,

$$\varepsilon(k) = \frac{\hbar^2k^2}{2m_e}. \tag{5.3}$$

[4] In insulators, these orbitals are completely filled such that the electrons cannot move.

[5] This model was developed by the German physicist Arnold Sommerfeld, who expanded an earlier theory developed by his compatriot Paul Drude. Hence it is formally known as the Sommerfeld–Drude model, but also goes by the names electron gas, Fermi gas, and Fermi sea models. Interestingly, Sommerfeld was nominated for the Nobel prize 84 times, but never won.

The corresponding electron momentum is

$$\vec{p} = \hbar\vec{k}. \tag{5.4}$$

There are two important aspects of this solution. First, like many types of partial differential equations, there are an infinite number of solutions associated with its eigenvalues. The eigenvalues, however, take discrete values. Accordingly, the eigenvalues of Schrödinger's equation lead to discrete values for the wave vector such as (in 1D),

$$k_x = \frac{2\pi n}{L_x}, \; n = 1, 2, 3..., \tag{5.5}$$

where L_x is the length of the crystal in the x-direction. Because of the Pauli exclusion principle, only two electrons (one spin up and the other spin down) may occupy any wave vector. Similarly, discrete electron wave vectors can be described for two-dimensional and three-dimensional systems.

With the wave vectors known, it is now possible to define the **density of states**. The density of states is the number of states N at each energy level ε available to be occupied. Conceptually, it is perhaps easier to think of a theater. The first energy level is ground floor level, and there are a certain number of seats or energy states N. The mezzanine is the second energy level with its own number of seats (states), followed by the first balcony, second balcony, and so forth. Mathematically, the density of states is

$$D(\varepsilon) = \frac{1}{V}\frac{dN}{d\varepsilon}, \tag{5.6}$$

and it can be shown that in 3D, the density of states for electrons is

$$D(\varepsilon) = \frac{1}{2\pi^2}\left(\frac{2m_e}{\hbar^2}\right)^{3/2}\sqrt{\varepsilon}. \tag{5.7}$$

Now equation (5.7) only represents the available energy states. However, what is still unknown is how many electrons are in those states, or continuing the theater analogy, equation (5.7) tells you how many seats are in each level of the theater but it is still unknown how many tickets have been sold for the seats. This is described by the statistics of the electrons.

5.2.2 Fermi–Dirac distribution and the work function

Like the Maxwell–Boltzmann energy distributions for particles in a gas (chapter 2), the electrons in a metal will have an equilibrium energy distribution that predicts the probability of an electron being at a certain energy ε. Because the underlying statistics are different, electrons do not follow the Maxwell–Boltzmann distribution, but instead follow the Fermi–Dirac distribution

$$f(\varepsilon) = \frac{1}{\exp\left(\frac{\varepsilon - \mu_c}{k_B T}\right) + 1}, \tag{5.8}$$

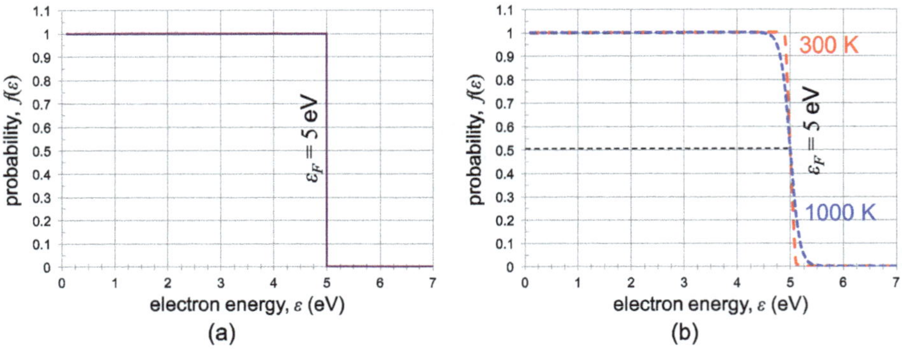

Figure 5.2. Plots of the Fermi–Dirac distribution function at (a) $T = 0$ K and (b) $T = 300$ K and $T = 1000$ K at a Fermi energy of $\varepsilon_F = 5$ eV.

where μ_c is the chemical potential of the metal[6], and T is the temperature of the metal. At a temperature of 0 K, the distribution is shown in figure 5.2(a), and it is clear that every level up to some maximum energy is fully occupied. This maximum energy is called the **Fermi energy** ε_F and is defined as the maximum energy an electron occupies at $T = 0$ K via

$$N_e = \int_0^{\varepsilon_F} D(\varepsilon)d\varepsilon, \tag{5.9}$$

where N_e is the number density of free electrons in the metal (which can be determined by the atomic structure and lattice geometry in a given volume). This leads to the Fermi energy being expressed as

$$\varepsilon_F = \frac{\hbar^2}{2m_e}(3\pi^2 N_e)^{2/3}. \tag{5.10}$$

As the temperature increases, the energy of the electrons increases and a few of the electrons near the Fermi energy are energized to a higher energy state. All the low-energy states are still occupied, but the occupation probability of the higher energy states is now greater than 0, as shown in figure 5.2(b)[7].

5.3 General electron emission

It is clear that as thermal energy is added to a metal, the free electrons are energized to higher energy levels (figure 5.2). At some energy, the high-energy electrons will have sufficient energy to actually 'break' away from the surface. The minimum energy (above the Fermi energy) required to remove an electron from the surface of

[6] Note that generally the chemical potential is approximately equal to the Fermi energy, $\mu_c \sim \varepsilon_F$.

[7] For any temperature above 0 K, the Fermi–Dirac distribution dictates that the probability of occupation at the Fermi energy is exactly 0.5. That is $f(\varepsilon_F) = 0.5$ for $T > 0$ K, and the distribution adjusts around this point as more higher energy states become occupied.

a solid to the vacuum outside the surface is called the work function, ϕ.[8] Therefore, generally, an electron emits when it has an energy greater than the work function. This is shown in figure 5.3, which is an electron motive diagram. The vertical axis is energy and the left of this axis represents the solid while the right represents the gas or vacuum outside the solid surface. The work function is the barrier potential energy and appears as a 'step' into the vacuum level. Also to the left of the axis is the Fermi–Dirac distribution of the electrons in the metal at some $T > 0$ K (notice it is rotated 90 degrees from the ones shown in figure (5.2)). It is clear that only a small number of electrons occupy high-energy states that are above the work function, and thus only a small number will be emitted. Electrons can be energized to occupy these high-energy states in different ways—by adding heat (**thermionic emission**), by absorbing the energy of a photon (**photoemission** or the photoelectric effect), from an electric field (**field emission**), or from an ion or excited metastable impacting the surface (**secondary emission**). This last one, secondary emission, is the critical one for gas discharges, and it is called the γ-process. However, a rudimentary understanding of the other three (and why they may or may not be important) is essential as well. Therefore, we will first consider thermionic, photo, and field emission and then finish with a detailed discussion of secondary emission.

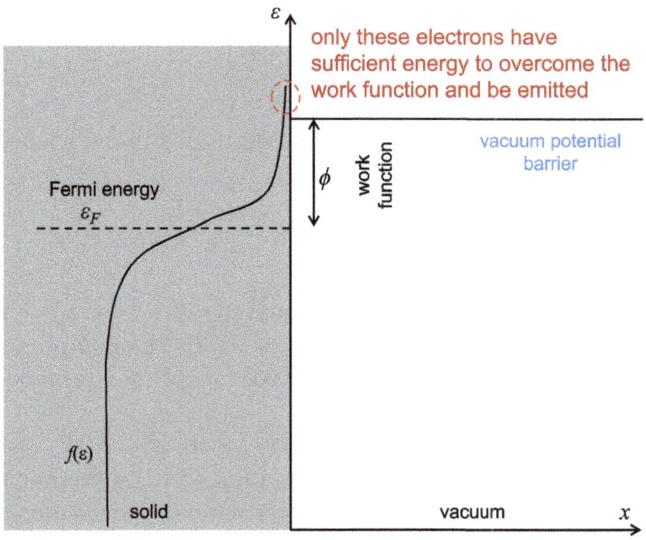

Figure 5.3. An electromotive diagram illustrating how only a small electron population has sufficient energy to overcome the potential barrier (work function) to be emitted into the vacuum.

[8] As it is an energy, the units of the work function are eV. Typical values for the work function are ~3–6 eV for most metals.

5.3.1 Thermionic emission

In thermionic emission the metal is heated to such an extent that the electrons pass over the work function potential barrier, as shown in figure 5.3. Generally, the emission current density j can be determined using

$$j = q\int_0^\infty N(\varepsilon)\mathscr{T}(\varepsilon)\mathrm{d}\varepsilon, \tag{5.11}$$

where $N(\varepsilon)$ is the *supply function* and depends on the density of states and Fermi–Dirac statistics, and $\mathscr{T}(\varepsilon)$ is the transmission coefficient and is determined by solving Schrödinger's equation for the wavefunction of the electron in the solid and in the vacuum. Conceptually, this can be thought of as the following: the current density depends on the number of available electrons at an energy level ($N(\varepsilon)$) that make it out of the solid ($\mathscr{T}(\varepsilon)$), and by integrating over all energies, the total current is determined. If we start with a relatively simple approximation, it is clear that the probability of emission is zero $\mathscr{T}(\varepsilon) = 0$ if the electron's energy is below the work function ($\varepsilon < \phi$) and that the probability of emission is one $\mathscr{T}(\varepsilon) = 1$ if the electron's energy is above the work function ($\varepsilon > \phi$). However, a more detailed analysis of the transmission coefficient can be used, and in the case of thermionic emission, this leads to the Richardson–Dushman equation[9]

$$j = AT^2 \exp\left(-\frac{\phi}{k_\mathrm{B}T}\right), \tag{5.12}$$

where T is temperature of the cathode in kelvins and A is the Richardson constant given by

$$A = \frac{4\pi m_e k_\mathrm{B} q}{h^3}. \tag{5.13}$$

By this definition, A is a universal constant equal to $1.201\,73 \times 10^6$ A m^{-2} K^{-2}. Note that the exponential relationship is a function of $-1/T$, which implies that in order to obtain any appreciable current, the cathode must be heated to very high temperatures.

Consider an order of magnitude calculation where $\phi \sim 5$ eV, $k_\mathrm{B} \approx 8 \times 10^{-5}$ eV K^{-1}, and $T = 300$ K. Plugging these into equation (5.12), one obtains $j \sim 10^{-80}$ A m^{-2}. This is an extremely small value, and any reasonably sized electrode would not achieve any emission current. On the other hand, if the electrode temperature is 2000 K, then $j \sim 10^{-1}$ A m^{-2}, which would correspond to an appreciable current for any reasonably sized electrode. This illustrates the impact of the exponential temperature dependence on the resulting current. Therefore, thermionic emission is only significant in high-temperature (thermal) plasmas or when there is appreciable

[9] Named after Englishman Owen Willans Richardson, who won a Nobel Prize in 1928 for his work on thermionic emission, and Russian Saul Dushman, who derived the general form and the form of the constant.

cathode heating such as in arcs[10]. However, in low-temperature discharges where the gas and ion temperature is ~300 K thermionic emission is not significant at all.

5.3.2 Photoemission

In photoemission (also called the photoelectric effect[11] or photoelectric emission), the solid absorbs incident photons (from a laser, for example) and these photons excite electrons to higher energy levels. If the electron is excited to an energy greater than the work function, then it can be emitted from the solid. The 'classical' relation is that the kinetic energy of the ejected electron will be

$$\varepsilon_e = h\nu - \phi. \tag{5.14}$$

From this concept, one can derive the energy distribution and therefore current of electrons [3]. Within the context of a discharge, it is important to consider the following two issues:

- The emission current's relation to the intensity of the bombarding light.
- The emission current's relation to the energy (frequency or wavelength) of the bombarding light.

Light intensity can be loosely defined as the rate of photons bombarding a surface. If the light is monochromatic (a single wavelength) all photons have the same energy and therefore the same probability of emitting an electron. Therefore, photoemission current is linearly proportional to the light intensity. The emission current increases with the frequency (or inversely with wavelength) because the greater the energy of the incoming light the greater the probability that an electron is emitted. Figure 5.4 shows these relationships qualitatively.

The threshold frequency for photoemission to occur is roughly

$$\nu_0 = \frac{\phi}{h}. \tag{5.15}$$

Consider an order of magnitude approximation where $\phi \sim 5$ eV, $h \approx 4 \times 10^{-15}$ eV s, and $c \approx 10^8$ m s^{-1}. Using equation (5.15), $\nu_0 \sim 10^{15}$ Hz, which corresponds to $\lambda_0 = \frac{c}{\nu_0} \sim 100$ nm. This wavelength corresponds to the ultraviolet portion of the electromagnetic spectrum. Because of the relaxation of excited neutrals to their ground state, photons with these wavelengths can be emitted in a discharge. Therefore, depending on the exact cathode material (or ϕ) and the intensity of the emission in the discharge, photoemission can be a source of electrons in a discharge.

[10] This is exactly the case in arcs, which often form localized ion channels called cathode spots that become extremely hot. In fact in arcs, as used in arc welding, thermionic emission is an essential charge producing mechanism [4].

[11] The photoelectric effect was discovered in 1887 by the German physicist Heinrich Hertz, and Albert Einstein won the Nobel Prize in 1921 for his work on the photoelectric effect, one of the four seminal papers he published in 1905.

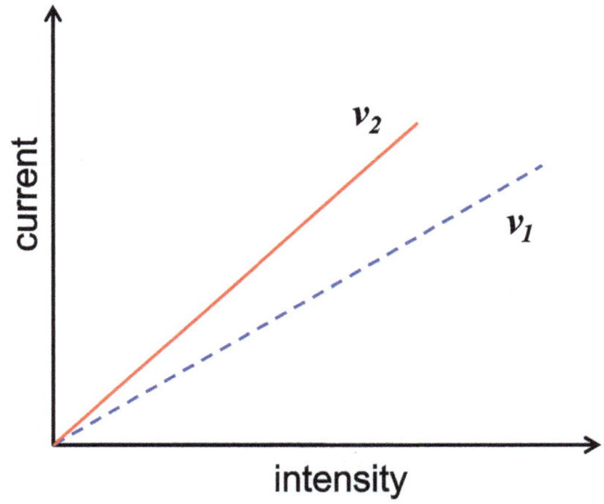

Figure 5.4. Qualitative relationship between current and light intensity for two different frequencies ($\nu_1 < \nu_2$ or $\lambda_1 > \lambda_2$).

5.3.3 Field emission

Field emission is different than thermionic and photoemission in that the electron in the cathode does not have to overcome the work function to escape from the cathode. In the presence of a very strong electric field, the potential barrier is both lowered and thinned; that is, rather than the 'step' potential barrier shown in figure 5.3, the barrier is more triangular, as shown in figure 5.5[12]. In this case, the electrons can quantum mechanically tunnel *through* the barrier rather than passing *over* the barrier.

To determine the field emission current density, equation (5.11) is again used, but the transmission coefficient is very different. Using an approximate solution for Schrödinger's equation, the Fowler–Nordheim equation, which takes a form similar to the Richardson–Dushman equation, is used to describe field emission[13]

$$j = \frac{A_{FN}E^2}{\phi t^2(y)} \exp\left(-\frac{B_{FN}\phi^{3/2}v(y)}{E}\right), \tag{5.16}$$

where A_{FN} and B_{FN} are constants ($A_{FN} = 1.54 \times 10^{-6}$ A m^2, $B_{FN} = 6.83 \times 10^7$), and ϕ is in eV and E in V m^{-1}. The variable y can be approximated as

$$y = 3.79 \times 10^{-4}\frac{\sqrt{E}}{\phi}, \tag{5.17}$$

[12] Any charge, positive or negative, generates an electric potential and field. In this instance, because electric potential can be summed using superposition, we must consider both the potential due to the electron just being emitted from the surface and the potential due to the applied voltage. The applied voltage produces the triangular shape while the 'image' charge of the emitted electron produces the rounded tip.

[13] The Fowler–Nordheim equation is named after the Englishman Ralph Howard Fowler and the German Lothar Wolfgang Nordheim, who published their seminal work in 1928 [5]. In 1956, Murphy and Good developed one of the most complete mathematical descriptions [6].

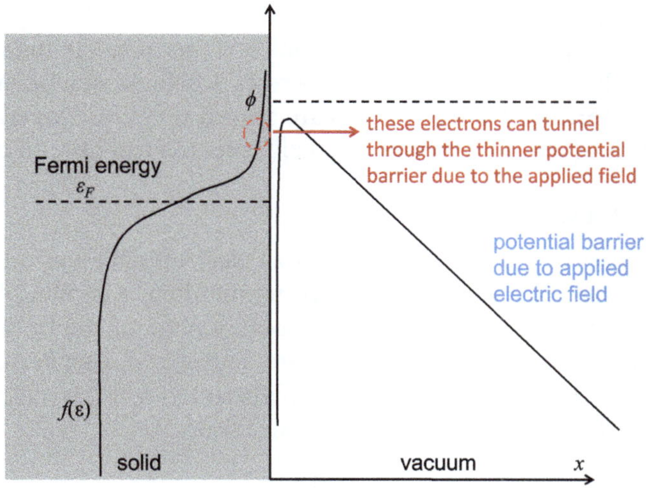

Figure 5.5. An electromotive diagram illustrating how an applied electric field thins the potential barrier so that electrons in energy states below the work function can tunnel out of the surface and be emitted into the vacuum. The potential barrier has two features—a linear portion due to the linear function of potential between two electrodes and a curved tip due to the 'image' potential of an electron just escaping the surface.

and the functions $t^2(y)$ and $v(y)$ can be approximated as

$$t^2(y) \approx 1, \tag{5.18}$$

$$v(y) \approx 0.95 - y^2. \tag{5.19}$$

If equation (5.16) was considered by itself, an electric field ~10^9 V m^{-1} would be required to generate any appreciable field emission current. However, the necessary electric field has been observed to be closer to 10^8 V m^{-1} in practice and, in special cases, as low as 10^7 V m^{-1} [7]. This is because of geometric field enhancement. At the surface of most electrodes, there will be microprotrusions and surface roughness where electric potential lines are distorted, increasing the local electric field at the cathode surface. This effect is typically included through a factor called β, and the current density equation can be rewritten as

$$j = \frac{A_{\mathrm{FN}}(\beta E)^2}{\phi t^2(y)} \exp\left(-\frac{B_{\mathrm{FN}}\phi^{3/2}v(y)}{\beta E}\right), \tag{5.20}$$

where

$$y = 3.79 \times 10^{-4}\frac{\sqrt{\beta E}}{\phi}. \tag{5.21}$$

For a typical metal, $\beta \sim 50$–100, whereas for specialized materials, such as carbon nanotubes or diamond electrodes, it has been reported to be as high as $\beta \sim 1000$ [7].

In general it can be concluded that field emission is largely negligible for discharges at any appreciable pressure (>1 mTorr), because the required voltage is simply much too high and other surface processes dominate well before that voltage is reached. However, when the electrodes are spaced very close together (~5 μm) [8] or the discharge is at very low pressure [9], field emission can play an important role.

5.4 Secondary emission

Secondary emission can be broadly defined as electron emission due to a particle (ion, neutral, metastable neutral, electron) bombarding a surface. In some texts, photoemission is called a secondary process because photons can be characterized as a particle. In this book the two processes are conceptually separated. In general, secondary emission is characterized by the coefficient γ, which is simply the ratio of emitted electrons to the ratio of incident particles, or

$$\gamma = \frac{N_{\text{emit}}}{N_{\text{incident}}}. \tag{5.22}$$

Based on this definition, γ can also be thought of as an emission probability.

The dominant secondary emission process in discharges is ion-induced emission, when the particles bombarding the cathode are positive ions that were formed in the discharge and are accelerated by the electric field toward the cathode (as shown in figure 5.1). A second, generally less important (although not always negligible) secondary emission process is metastable-induced emission, when the bombarding particles are metastables (long-lived excited neutrals) that were formed in the discharge and slowly diffuse toward the cathode. (Being neutrals, they are not accelerated by the electric field.) Other types of secondary emission (electron, etc.) may occur, but only under special circumstances and will not be considered here.

5.4.1 Ion-induced secondary emission

When considering a positive ion accelerated toward the cathode, it is intuitive to believe that secondary emission occurs because the kinetic energy of the ion 'knocks' an electron out of the cathode. While this may occur under special cases, in general the secondary emission process is much more nuanced. There are, in fact, two ways the electron is ejected. One is the aforementioned kinetic energy process, but as will be shown this requires much greater energy than is typical for an ion in a discharge. The second is the less intuitive, but more prominent, potential energy process (also called the Auger process), which occurs quantum mechanically.

In the kinetic energy process, the bombarding ion approaches the cathode and through a 'binary' collision with an electron, imparts sufficient kinetic energy for the electron to overcome the work function ϕ of the metal and be emitted. The threshold velocity necessary to eject an electron can be estimated from this conceptual image.

Consider an electron of mass m_e with energy at the Fermi level ε_F, corresponding to a Fermi velocity of the electron of v_F.[14] The maximum energy exchange from a

[14] The Fermi energy can easily be converted into velocity via $v_F = \sqrt{2\varepsilon_F/m_e}$.

heavy ion ($m_i \gg m_e$) impacting the electrode with velocity v_i to an electron inside the electrode with velocity is v_F is

$$\Delta\varepsilon = 2m_e v_i(v_i + v_F). \qquad (5.23)$$

If $\Delta\varepsilon = \phi$, the minimum energy required for an electron at the Fermi level to be emitted, then equation (5.23) can be rearranged to determine the threshold incident ion velocity,

$$v_{i,\text{th}} = \frac{1}{2}\left[-v_F + \sqrt{\left(v_F^2 + \frac{2\phi}{m_e}\right)}\right]. \qquad (5.24)$$

Now consider an order of magnitude approximation where $\phi \sim 5$ eV and $\varepsilon_F = 7$ eV, corresponding to $v_F \approx 1.5 \times 10^6$ m s.[15] For an electron mass of $m_e \approx 9 \times 10^{-31}$ kg, equation (5.24) gives an ion threshold incident velocity of $v_{i,\text{th}} \sim 10^5$ m s^{-1}. For an air molecule, where $m_i \sim 10^{-26}$ kg, $\varepsilon_{i,\text{th}} = (1/2)m_i v_{i,\text{th}}^2 \sim 10^4$ eV. Compare this to the typical energy needed for an electron to ionize a neutral molecule (~10 eV), and it is clear that most ions will never approach an energy of 1000–10 000 eV in a typical discharge. In fact, in a low-temperature discharge, the average ion temperature is on the same order of magnitude as the gas temperature or six orders of magnitude lower than that necessary for kinetic secondary emission. Therefore, while there may be a small population of high-energy ions, none typically approach these types of energies, such that the kinetic process has a negligible contribution to ion-induced secondary emission.

The potential energy process is the dominant ion-induced secondary emission mechanism. As the 'slow' ion approaches the surface, it creates an electrical potential energy well as it comes within a few atomic radii of the surface (on the order of a de Broglie wavelength). The potential barrier is therefore only a few atomic radii wide, as well. An electron with some energy $\varepsilon_{e,\text{cath}}$ defined relative to the work function (and typically near the Fermi energy) tunnels out of the solid (similar to field emission), gets trapped in the potential well, and interacts with the ion. There are two possibilities: neutralization and Auger emission. These will be discussed separately.

Neutralization. In neutralization, the electron tunnels into one of the *excited* states of the positive ion to form a neutral gas molecule:

$$e^-_{\text{cath}} + M^+ \rightarrow M^*.$$

Both metastable and short-lived excited states are possible. The resultant energy state of this excited neutral (the former ion) is simply the difference between the ionization energy of the neutral and the energy gained from the electron that tunneled from the cathode,

$$E_{M^*} = E_{\text{ion}} - \varepsilon_{e,\text{cath}}.$$

[15] Note that these energies must be converted to joules for this calculation, where 1 eV = 1.602 18 $\times 10^{-19}$ J.

As with all excited states and metastables, the electron will relax down to a ground state, releasing a photon. This is called *recombination radiation* (or radiative recombination), or

$$M^{*} \rightarrow M + h\nu,$$

where the photon has an energy equal to the difference between the excited and ground states in the neutral

$$h\nu = E_{M^{*}} - E_{M}.$$

This process is illustrated qualitatively in figure 5.6. Neutralization is the most probable event that occurs as the ion approaches the cathode.

Auger emission. In Auger emission, the cathode electron tunnels into the ground state of the positive ion to form a neutral gas molecule. However, there will be an energy mismatch between the captured electron ($\varepsilon_{e1,\text{cath}}$) and the ground state of the neutral gas molecule (E_M), or

$$\Delta\varepsilon = \varepsilon_{e1,\text{ cath}} - E_{M}.$$

A second electron *in the cathode* absorbs this excess energy; if the energy excites this second electron to be above the work function, it will be emitted. That is, the second electron in the cathode absorbs the excess energy and is promoted to a high-energy state. If we assume the second electron is at an energy of $\varepsilon_{e2,\text{ cath}}$, quantitatively, there will be emission if

$$\Delta\varepsilon > \varepsilon_{e2,\text{cath}},$$

recalling that $\varepsilon_{e2,\text{cath}}$ is defined relative to the work function. We can write this reaction as

$$e^{-}_{1,\text{cath}} + e^{-}_{2,\text{cath}} + M^{+} \rightarrow e^{-}_{2} + M.$$

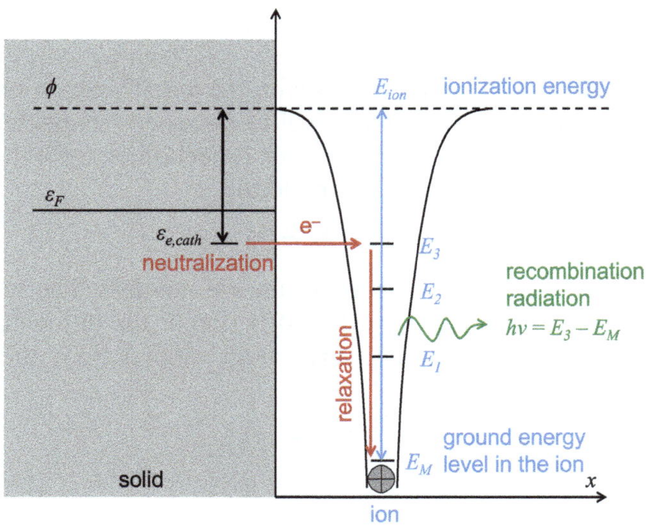

Figure 5.6. Illustration of the neutralization process where an electron tunnels from the cathode into the excited energy level E_3 of the ion/neutral.

The energy of the ejected electron will clearly be

$$\varepsilon_e = \Delta\varepsilon - \varepsilon_{e2,\mathrm{cath}} = (\varepsilon_{e1,\mathrm{cath}} - E_M) - \varepsilon_{e2,\mathrm{cath}}.$$

Typically, we set the vacuum level as the reference point, as shown in figures 5.6 and 5.7. In this case, we can rewrite the excess energy as

$$\Delta\varepsilon = E_{\mathrm{ion}} - \varepsilon_{e1,\mathrm{cath}},$$

and we can consider the energy required to cause Auger emission as

$$E_{\mathrm{ion}} - \varepsilon_{e1,\mathrm{cath}} > \varepsilon_{e2,\mathrm{cath}}.$$

The ejected electron has an energy of

$$\varepsilon_e = \Delta\varepsilon - \varepsilon_{e2,\mathrm{cath}} = E_{\mathrm{ion}} - \varepsilon_{e1,\mathrm{cath}} - \varepsilon_{e2,\mathrm{cath}}.$$

In summary, we can consider four possibilities for what can happen when an ion approaches the cathode surface:

1. *Neutralization-radiation.* The ion is neutralized by an electron tunneling into an excited state, and neutral relaxes into its ground state, releasing a photon through recombination radiation. (Highly probable.)
2. *Neutralization.* The ion is neutralized by an electron tunneling into its ground state and imparts energy on a second electron in the cathode where $\Delta\varepsilon = E_{\mathrm{ion}} - \varepsilon_{e1,\,\mathrm{cath}} < \varepsilon_{e2,\,\mathrm{cath}}$, such that the second electron *is not* emitted. (Moderately probable.)
3. *Secondary emission by Auger emission.* The ion is neutralized by an electron tunneling into its ground state and imparts energy on a second electron in the

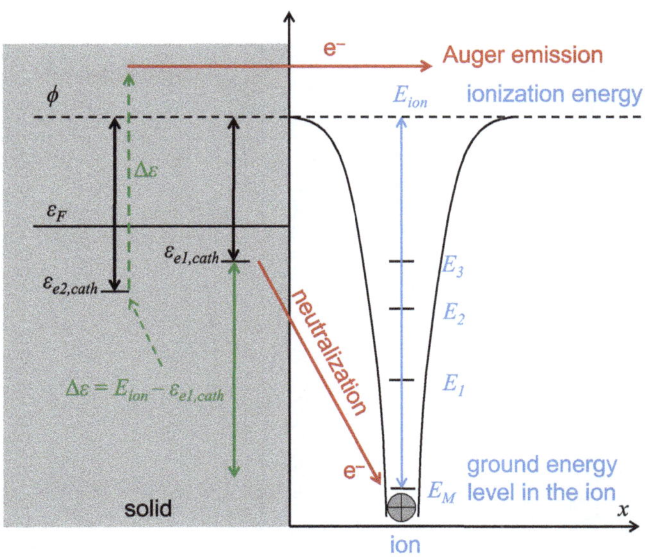

Figure 5.7. Illustration of the Auger emission process where an electron tunnels from the cathode into the ground state of the ion/neutral, and a second electron receives the excess energy and is emitted.

cathode where $\Delta\varepsilon = E_{\text{ion}} - \varepsilon_{e1,\,\text{cath}} > \varepsilon_{e2,\,\text{cath}}$ resulting in an emitted electron. (Less probable.)

4. *Secondary emission by kinetic energy.* The ion impacts the surface with sufficient kinetic energy to promote an electron above the work function resulting in an emitted electron. (Highly non-probable.)

Secondary emission, therefore, is essentially the Auger emission of electrons due to neutralization of ions at the electrode surface.

5.4.2 Secondary emission coefficient γ

Now let us consider approximations for ion-induced secondary emission and the secondary emission coefficient γ. The condition for Auger emission $\Delta\varepsilon > \varepsilon_{e2,\,\text{cath}}$ is most easily met when both $\varepsilon_{e1,\,\text{cath}} = \varepsilon_{e2,\,\text{cath}} = \phi$, as shown in figure 5.8. In this case, the condition for secondary emission is that the ionization energy must be greater than twice the work function

$$E_{\text{ion}} \geqslant 2\phi. \tag{5.25}$$

If this is the *easiest case*, then it is clear that the *maximum* kinetic energy of an emitted electron must be

$$\varepsilon_{e,\max} = E_{\text{ion}} - 2\phi. \tag{5.26}$$

For the potential energy process, unlike the kinetic energy process, there is no 'threshold velocity'; neutralization will occur even if there are low-speed ions. The driving requirement as to whether an Auger emission process occurs is whether the ionization energy of the gas neutral is more than twice that of the work function;

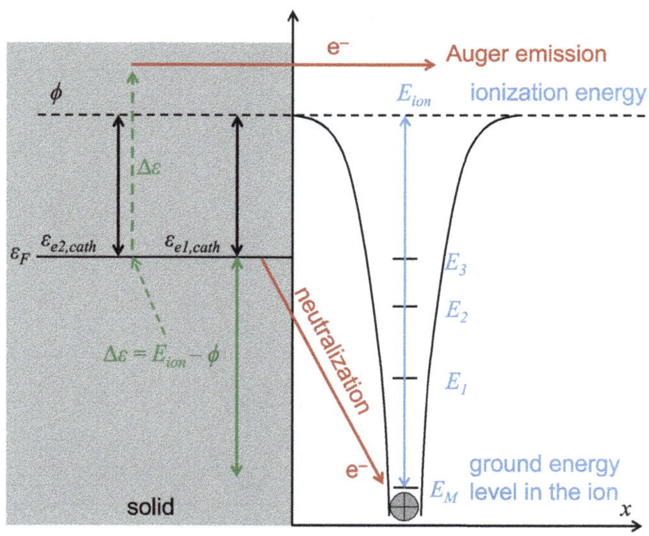

Figure 5.8. Illustration of the easiest Auger emission process where the two cathode electrons are at the Fermi energy.

that is, the gas molecule and electrode material chemistry must match[16]. What is a good match? For a gas, it is best to have a high ionization energy and for the cathode it is best to have a low work function. Or simply:

- The *gas* should have high ionization energy → noble gases (He > 25 eV, Ne > 20 eV, Ar, Kr, Xe, Rn).
- The *electrode* should have a low work function → alkali (Li ~ 2.93 eV, Na ~ 2.36 eV, K ~ 2.29 eV, Rb, Cs) or alkaline earth metals (Be, Mg, Ca, Sr, Ba, Ra).

Based on this limiting criterion, empirical relations for γ have been proposed, such as[17]

$$\gamma \approx 0.016(E_{\text{ion}} - 2\phi), \tag{5.27}$$

where the energy difference is in electronvolts. It essential to note that this is only an *approximation* and is accurate to only within 50% for clean metals. The reality is that surface conditions, morphology, impurities, and contamination play a very important role. In practice, γ typically ranges from 10^{-4} to 10^{-1}, which indicates that typically less than 10% and often less than 1% of the ions in the discharge produce secondary emitted electrons. However, as noted in section 5.1, this small supply of electrons is *essential* to discharge formation.

5.4.3 Metastable-induced secondary emission

Like an ion, a metastable can also induce an energy exchange process at the cathode that results in electron emission. Like Auger emission, it is a potential energy process because the metastable has excess energy and will eventually relax back to its ground state. For this reason it is fairly efficient at secondary emission. This is often called *Auger de-excitation-emission* (or *Auger relaxation-emission*). The energy of the ejected electron will now be

$$\varepsilon_e = E_{M^*} - \varepsilon_{e,\,\text{cath}}.$$

Again, like with Auger neutralization-emission, the 'best case scenario' is when the electron in the cathode solid is at the Fermi level or $\varepsilon_{e,\,\text{cath}} = \phi$. In this case, the condition for secondary emission is that the excitation energy of the gas molecule be

$$E_{M^*} \geqslant \phi.$$

In general, the secondary emission coefficient for metastable emission γ_{M^*} can range from 10^{-2} to 10^{-1}. Note that is at least equal to if not greater than the secondary emission coefficient due to ions. However, the primary difference between a metastable gas particle and a gas ion is that the metastable is neutral, which means that it is not accelerated toward the cathode by an electric field. A large number of metastables will be lost by

---

[16] This is not entirely true, there is some dependence not only on the speed of the ion but also its angle of impact [10, 11], but at a basic level these can be neglected.

[17] See, for instance, [12].

relaxing to their ground state before reaching the cathode or by diffusing to the anode or to other walls/boundaries instead of diffusing to the cathode. Thus in many cases, metastable secondary emission is negligible compared to ion secondary emission not because the emission process is less efficient but because the metastable flux is much lower than the ion flux to the cathode.

5.5 Summary

The purpose of this chapter was to give an overview of the creation of electrons from the cathode that are critical charge carriers and thus help in the development of a discharge. The following processes were discussed:

- The ion-induced secondary emission process is dominant, and it is an *Auger potential energy process* rather than a *kinetic energy process.* It is described by the coefficient γ.
- The metastable-induced secondary emission process can also occur, but to a lesser extent, and it is also an Auger potential energy process.
- Photoemission (the photoelectric effect) can be important depending on the gas (producer of photons) and cathode material (work function), but is also generally less important when compared to secondary emission.
- Thermionic emission is typically not important unless the cathode is at high temperature (such as in equilibrium discharges like arcs).
- Field emission is typically not important except in some special cases (such as in extremely small gaps or at very high vacuum).

It is important to remember that more than one of these emission phenomena can be active in a given discharge scenario, and it is virtually impossible to identify exactly how each electron is individually produced. In practice, it is difficult to distinguish between the ion, metastable, and photoemission electrode processes that may all be occurring. Therefore, we generally collect them all into a single, effective coefficient that we call γ with a typical value ~10^{-2}–10^{-1}.

References

[1] Ashcroft N W and Mermin N D 1976 *Solid State Physics* (Boston, MA: Thomson Learning)

[2] Kittel C 2004 *Introduction to Solid State Physics* 8th edn (Weinheim: Wiley)

[3] Dubridge L A 1933 Theory of the energy distribution of photoelectrons *Phys. Rev.* **43** 727–41

[4] Anders A 2008 *Cathodic Arcs: From Fractal Spots to Energetic Condensation* (New York: Springer)

[5] Fowler R H and Nordheim L 1928 Electron emission in intense electric fields *Proc. R. Soc. Lond.* A **119** 173–81

[6] Murphy E L and Good R H 1956 Thermionic emission, field emission, and the transition region *Phys. Rev.* **102** 1464–73

[7] Bonard J-M, Salvetat J-P, Stöckli T, Forro L and Chatelain A 1999 Field emission from carbon nanotubes: perspectives for applications and clues to the emission mechanism *Appl. Phys.* A **69** 245–54

[8] Go D B and Venkattraman A 2014 Microscale gas breakdown: ion-enhanced field emission and the modified Paschen's curve *J. Phys. D: Appl. Phys.* **47** 503001

[9] Fursey G N 1985 Field emission and vacuum breakdown *IEEE Trans. Elect. Insul.* **EI-20** 659–70

[10] Phelps A V and Petrovic Z L 1999 Cold-cathode discharges and breakdown in argon: surface and gas phase production of secondary electrons *Plasma Sources Sci. Technol.* **8** R21–4

[11] Phelps A V, Pitchford L C, Pedoussat C and Donkó Z 1999 Use of secondary-electron yields determined from breakdown data in cathode-fall models for Ar *Plasma Sources Sci. Technol.* **8** B1–2

[12] Raizer Y P 1991 *Gas Discharge Physics* (Berlin: Springer)

Part II

Ion transport

IOP Publishing

Ionization and Ion Transport (Second Edition)
A primer for the study of gas discharges and plasmas
David B Go

Chapter 6

Foundations from general transport theory

Now that we have established how ions and electrons are *generated* in a gas discharge, we have to establish a clearer picture of how ions and electrons (and neutrals for that matter) *move* in a discharge. This is in fact very closely tied to ionization since the primary ionization process, the α-process, is a *collision* process, thus relying on how the particles move and interact. This part of the book is separated into two chapters. Chapter 6 deals with foundational transport concepts, moving from the molecular view of transport to the continuum scale. Chapter 7 then applies these concepts to gas discharges in general, deriving the governing equations that describe the gas discharges and fundamental scaling that arise from these descriptions.

We first discussed transport concepts in chapter 2, when we introduced the concept of a collision between two gas particles, the collision cross-section σ (which relates to the probability of a collision), and the mean free path $\bar{l}$. All of these, obviously, are tied to the *transport* of gas particles, and without these collisions we would not see energy distributions, such as the Maxwell-Boltzmann distribution, for either gas particles or electrons[1]. However, we are going to move one step further and consider how these collision concepts tie to 'macroscopic' transport behavior. We are also going to look at the basic equations that govern all transport; equations that we will formally apply to gas discharges in chapter 7.

6.1 Basics of transport

First we must explain what we mean by 'transport'[2]. There are two ways of thinking about it. First of all, there is simply the transport of the particles themselves. In the

[1] That is, without collisions, there is no redistribution of energy.

[2] For a good overview of general, widely applicable transport concepts, see *Analysis of Transport Phenomena* by Deen [2].

absence of an external force on a collection of gas particles[3], they will naturally spread until they are equally distributed everywhere, a process called *diffusion*. In this case, we are describing self-diffusion, but you can also have diffusion of one type of particle into a medium of another type of particle[4]. Conceptually, since the particles want to be equally distributed everywhere, they naturally will move from areas of high concentration to low concentration. That is, they are driven by the concentration *gradient*. The second way of thinking about transport is that as the particles move, they carry with them some property, such as momentum or energy. That is, if we have a region where the particles have large momenta and kinetic energies, they will spread this momentum and energy by colliding with the particles around them. Thus the momentum and energy are *transported*, and similarly to diffusion, they rely on a *gradient* of momentum and/or kinetic energy to spread[5].

We can describe these basic transport concepts through what are known as *constitutive equations* that relate the flux of a property to the gradient of the property. For example, **Fick's law**[6] relates the flux J_N ($\mathrm{m^{-2}\,s^{-1}}$) of particles, where N is the particle density, to the gradient in the particle density or

$$J_N = -D\frac{\partial N}{\partial x}. \tag{6.1}$$

The parameter D is the *diffusion coefficient* or *diffusivity* ($\mathrm{m^2\,s^{-1}}$), and is simply a constant of proportionality between the flux and the gradient. Similarly, if momentum is essentially the velocity of the particles (neglecting mass for the time being), then the flux of momentum τ_s is related to the speed of the particles v by **Newton's law**[7]

$$\tau_s = -\mu\frac{\partial v}{\partial y}. \tag{6.2}$$

Conventionally, τ_s is known as the shear stress ($\mathrm{N\,m^{-2}}$) and μ is called the *dynamic viscosity* ($\mathrm{N\,s\,m^{-2}}$), and again is simply a constant of proportionality. Finally, the flux of energy, called the heat flux q ($\mathrm{W\,m^{-2}}$), depends on the energy gradient. Since energy can be related to temperature, we can write **Fourier's law**[8] as

$$q'' = -\kappa\frac{\partial T}{\partial x}. \tag{6.3}$$

The constant of proportionality κ is called the *thermal conductivity* ($\mathrm{W\,m^{-2}\,K^{-1}}$).

[3] For example, in a piston-cylinder, the wall of the piston exerts a pressure on the gas. In a gas discharge, the applied electric field exerts a Coulombic force on the charged particles.

[4] Conceptually, one might consider the diffusion of perfume or cologne sprayed in one corner of a room throughout the room.

[5] Clearly if there is no gradient, then momentum and energy will not spread as there is no place to spread them. Therefore, although momentum and energy may be exchanged during individual particle–particle collisions, from a macroscopic perspective the momentum and energy are 'equal everywhere'.

[6] Named after German physiologist Adolph Fick, who made pioneering contributions to medicine and physiology by applying a rigorous perspective of physics to the fields.

[7] Named after famed Englishman Sir Isaac Newton, one of the most prominent scientists in history.

[8] Named after French physicist and mathematician Jean-Baptiste Joseph Fourier, who is one of the most prominent contributors to the field of heat transfer and is often credited with discovering the greenhouse effect.

6.1.1 Application of kinetics to transport

While we can describe the relationship between the flux and gradient phenomenologically, the constants of proportionality (i.e. D, μ, and κ) need to be tied to the physical properties of the system. Luckily, for a gas, we determine these based on basic kinetic concepts. Consider a volume of particles that are moving randomly in every direction. If we draw a plane at position x, we can consider the flux of particles moving across this position x that *carry* with them some *scalar* property, ϕ.[9] We assume that particles passing through the plane from the left to the right move through a distance dx without any collisions and similarly particles going right to left also move through a distance dx from the plane without collisions, as shown in figure 6.1.

The *net* flux of particles, and hence scalar ϕ, passing through plane x is simply the difference between those moving left to right and those moving right to left or

$$J = \frac{1}{2}(N\phi v_x)_{x-dx} - \frac{1}{2}(N\phi v_x)_{x+dx}, \tag{6.4}$$

where N is the number density of particles, and v_x is the x-component of their velocity. The factor of $\frac{1}{2}$ assumes that only half of the particles in the distance dx move in the positive direction. If we apply a Taylor series expansion to the quantities around $+dx$ and $-dx$, respectively, we can show that the net flux is

$$J = -\frac{\partial(N\phi v_x)}{\partial x}dx. \tag{6.5}$$

The quantity dx is equivalently the x-velocity multiplied by the average time between collisions[10] τ so that equation (6.5) can be rewritten as

$$J = -v_x\tau\frac{\partial(N\phi v_x)}{\partial x}. \tag{6.6}$$

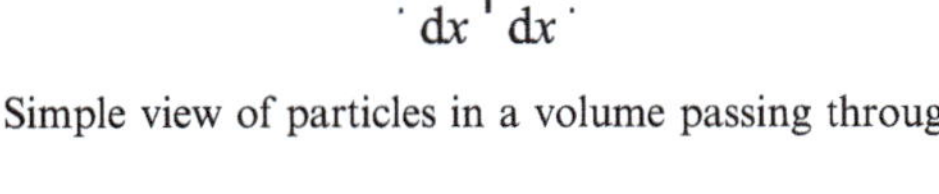

Figure 6.1. Simple view of particles in a volume passing through a plane x.

[9] For example, if the particles are ions they carry the scalar properties charge and mass. If they are neutral particles, they carry mass alone.
[10] The time analog of the mean free path is the mean collision time.

If we assume that v_x is not dependent on position and thus not a function of x and that it is equal to 1/3 of the total average speed, $v_x^2 = \frac{1}{3}\bar{v}^2$, this becomes

$$J = -\frac{1}{3}\bar{v}^2\tau\frac{\partial(N\phi)}{\partial x}. \tag{6.7}$$

We can relate equation (6.7) to our constitutive relationships in section 6.1 by how we define the scalar ϕ. For example, if we are only interested in the diffusion of the particles themselves, then they carry no scalar property and $\phi = 1$. If we equate, equation (6.1), Fick's law, to the generic definition of flux in equation (6.7), we see that

$$J_N = -D\frac{\partial N}{\partial x} = -\frac{1}{3}\bar{v}^2\tau\frac{\partial N}{\partial x}, \tag{6.8}$$

and that the diffusivity is equivalently $D = \frac{1}{3}\bar{v}^2\tau$. Similarly, if we consider the scalar property to be the x-component of momentum, mv_x, we can equate equation (6.2), Newton's law, to the generic definition of flux in equation (6.7) and show that dynamic viscosity is $\mu = \frac{1}{3}\bar{v}^2\tau Nm$ or, defining the mass density as $\rho = Nm$, $\mu = \frac{1}{3}\bar{v}^2\tau\rho$. Finally, we can apply a similar exercise to internal energy u and Fourier's law in equation (6.3) to show that the thermal conductivity is $\kappa = \frac{1}{3}\bar{v}^2\tau\rho c$, where c is the specific heat of the gas.

While these three transport properties appear distinct, they are in fact intimately related. First, all of these properties depend on the quantity $\bar{v}^2\tau$. The mean time between collisions, often called the *relaxation time*, is related to the mean free path $\bar{l}$ via $\tau = \bar{l}/\bar{v}$. As discussed in chapters 2 and 3, we can obtain approximations for the average speed and mean free path depending on the exact conditions of the gas, and thus the mean collision time as well. But even more interestingly is that, while $D = \frac{1}{3}\bar{v}^2\tau$, if we divide the dynamic viscosity by ρ we *also* obtain $\mu/\rho = \frac{1}{3}\bar{v}^2\tau$, and if we divide the thermal conductivity by ρc, we obtain $\kappa/(\rho c) = \frac{1}{3}\bar{v}^2\tau$. That is, when properly normalized (viscosity by mass and thermal conductivity by thermal mass), the diffusion properties of number density, momentum, and energy have the same form (and the same units of $m^2\ s^{-1}$)! This illustrates that regardless of the scalar property, diffusion of anything is essentially the same process—particles spreading. The quantity $\nu = \mu/\rho$ is known as the *kinematic viscosity* and the quantity $\alpha = \kappa/(\rho c)$ is known as the *thermal diffusivity*. Bringing all of this together, table 6.1 summarizes

Table 6.1. Constitutive laws for the flux of various scalars and the approximations for the transport coefficients from kinetic theory.

Scalar (ϕ)	Flux equation	Transport coefficient
Particle density (N)	$J_N = -D\frac{\partial N}{\partial x}$	$D = \frac{1}{3}\bar{v}\bar{l}$
Momentum (mv)	$\tau = -\mu\frac{\partial v}{\partial y}$	$\mu = \frac{1}{3}\rho\bar{v}\bar{l}$
Energy (u)	$q'' = -k\frac{\partial T}{\partial x}$	$\kappa = \frac{1}{3}\rho c\bar{v}\bar{l}$

the primary constitutive relationships and estimates for the transport properties based on kinetic theory.

6.2 Transport equations

The kinetic equations for transport properties treat the transport at a particle level, but what is needed is an understanding of the transport of the collective—that is, the gas medium as a collection of particles. As we saw in chapter 2, the collection of particles can be described by a *distribution function*, $f(\varepsilon)$, and this distribution will depend on how the particles (and their properties momentum and energy) are transported. That is, the distribution can change in space (from position x to $x + \Delta x$) and in time (from t to $t + \Delta t$).

There are thus two primary ways to consider transport. One is that we actually look at the transport of the distribution itself, the evolution of the distribution in space and time—and that at each instance of space and time, we describe the main properties of the collection of particles (number density, momentum, and energy) based on the distribution. This is a kinetic perspective and requires solving the **Boltzmann transport equation** (BTE). Alternatively, we can simply consider the *average* of each property (number density, momentum, and energy) and how these change in space and time, without ever considering the distribution explicitly. This is the continuum perspective and requires solving some form of the **scalar transport equation**. Here, we will first discuss the continuum perspective and then the kinetic perspective, concluding with how these two are related. If these equations are clear, than the specific transport equations for gas discharges follow naturally.

6.2.1 Continuum perspective—scalar transport equation

The general transport equation or, more accurately, the *scalar transport equation* is a generalized conservation equation for the motion of a scalar quantity of some medium through a domain. This equation is a *continuum* equation because it does not treat individual particles but rather the bulk—that is, the scalar quantity at any given position and time is its average value for all the particles at that same given poistion and time. Consider, for example, a gas. When we describe the scalar quantity of speed, we do not imply that every atom in the gas is moving at the same speed. Some may be (and, in fact, are) moving at different speeds than their neighbors. However, on average, the gas is moving at speed $\bar{v}$. If the gas is at equilibrium, we found in chapter 2 that this average is based on the Maxwell–Boltzmann distribution.

Let us begin to think about transport generally from a control volume approach. Here we will only consider a one-dimensional (1D) Cartesian domain, but it is fairly straightforward to extend this to three dimensions (3D) and/or cylindrical or spherical coordinates. The net 'transport' of any scalar quantity, which we will call ϕ, is the balance of the influx of ϕ across the boundaries into the domain, the outflux of ϕ across the boundaries from the domain, the accumulation of ϕ in the domain, the generation of ϕ within the boundaries of the domain, and the consumption or loss of ϕ within the boundaries of the domain. These transport processes are illustrated in figure 6.2. We can write the general conservation equation as a balance of all these processes:

$$\phi_{\text{accum}} = \phi_{\text{in}} - \phi_{\text{out}} + \phi_{\text{gen}} - \phi_{\text{loss}}. \tag{6.9}$$

It is important to note that the amount of ϕ depends on the actual number of particles that are participating in each transport mechanism—that is, ϕ is a quantity that is explicitly tied to the amount of gas that flows into or out of the domain or is generated/consumed within the domain. This becomes clear when we consider each of these terms individually.

Accumulation: The accumulation of ϕ over a time increment Δt is

$$\phi_{\text{accum}} = (N\phi V)_{t+\Delta t} - (N\phi V)_t.$$

The amount of ϕ is multiplied by the total number of particles $N{\cdot}V$, where N is the number density of particles and V is the volume of the control volume. In three dimensions, the volume of the control volume is $V = dx{\cdot}dy{\cdot}dz$. However, in 1D, dy and dz can be considered to be 1, so we can rewrite the accumulation term as

$$\phi_{\text{accum}} = (N\phi dx)_{t+\Delta t} - (N\phi dx)_t.$$

A Taylor expansion can be applied to the first term

$$(N\phi dx)_{t+\Delta t} \approx (N\phi dx)_t + \frac{\partial(N\phi dx)}{\partial t}dt$$

such that the accumulation term now becomes

$$\phi_{\text{accum}} \approx (N\phi dx)_t + \frac{\partial(N\phi dx)}{\partial t}dt - (N\phi dx)_t = \frac{\partial(N\phi)}{\partial t}dxdt.$$

Note that this assumes that the volume dx is invariant with time.

Influx and outflux: The influx and outflux terms can be considered together as

$$\phi_{\text{in}} - \phi_{\text{out}} = (J_\phi dt)_x - (J_\phi dt)_{x+\Delta x},$$

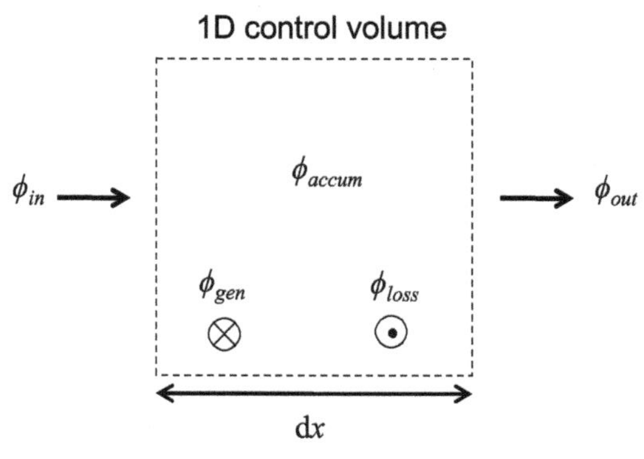

Figure 6.2. Basic control volume for the transport of a scalar quantity.

where J_ϕ is the flux of ϕ in the x-direction. Again, a Taylor expansion can be applied,

$$(J_\phi dt)_{x+\Delta x} \approx (J_\phi dt)_x + \frac{\partial(J_\phi dt)}{\partial x}dx,$$

such that the flux balance terms now become

$$\phi_{\text{in}} - \phi_{\text{out}} \approx (J_\phi dt)_x - \left[(J_\phi dt)_x + \frac{\partial(J_\phi dt)}{\partial x}dx\right] = -\frac{\partial(J_\phi)}{\partial x}dxdt.$$

Generation and loss: The generation and loss terms, often called source and sink, are conceptually similar and thus can be considered together as

$$\phi_{\text{gen}} = S_{\text{gen}}dxdt$$
$$\phi_{\text{loss}} = S_{\text{loss}}dxdt,$$

where S is the generation/loss rate of ϕ per unit volume. In this case, there is no need for a Taylor expansion to simplify the terms.

With these approximations for each term, equation (6.9) can be rewritten as

$$\frac{\partial(N\phi)}{\partial t}dxdt = -\frac{\partial(J_\phi)}{\partial x}dxdt + S_{\text{gen}}dxdt - S_{\text{loss}}dxdt. \tag{6.10}$$

Canceling the dx and dt terms and moving the flux term to the other side, equation (6.10) reduces to the general scalar transport equation:

$$\frac{\partial(N\phi)}{\partial t} + \frac{\partial(J_\phi)}{\partial x} = S_{\text{gen}} - S_{\text{loss}}. \tag{6.11}$$

It should be clear that this can be easily expanded into three dimensions as

$$\frac{\partial(N\phi)}{\partial t} + \vec{\nabla}\cdot\vec{J}_\phi = S_\phi, \tag{6.12}$$

where we have combined the generation and loss of ϕ into a single term S_ϕ for simplicity.

The only term that is not well defined in equation (6.12) is the flux vector $\vec{J}$. This term actually can represent a variety of physics, depending on what is driving the *flux* of the scalar. Here, we will lay out two common processes that comprise the flux vector: diffusion and advection[11]. *Diffusion* is the transport of the scalar quantity due to spreading, as discussed in section 6.1. *Advection* is the transport of the scalar quantity due to the *bulk motion* of a fluid, i.e. the fluid carries the scalar quantity as the fluid itself moves. These terms can be written as:

[11] Advection is often also called convection, but this can be confusing as thermal convection in heat transfer has both advective and diffusive components. Here we use the term advection to isolate the behavior from diffusion.

Diffusion: $\vec{J}_{\text{diff}} = -D_\phi \vec{\nabla} N\phi$, where D_ϕ is the diffusivity of the scalar ϕ.
Advection: $\vec{J}_{\text{adv}} = N\phi\vec{v}$, where $\vec{v}$ is the bulk velocity of the fluid.

Plugging these descriptions for diffusion and advection into equation (6.12) and with some rearranging produces a general **advection-diffusion equation:**

$$\frac{\partial(N\phi)}{\partial t} + \vec{\nabla}\cdot(N\phi\vec{v}) = \vec{\nabla}\cdot\left(D_\phi \vec{\nabla}(\,N\phi)\right) + S_\phi. \tag{6.13}$$

As written here, this is called a *conservative* form since all quantities are explicitly conserved. There are mathematical manipulations that can be used to change this form of the equation and put it into *non-conservative* form as well, but those will not be considered here.

As discussed in section 6.1, the nature of the diffusion depends on the scalar ϕ in question. If the scalar is simply the number of particles (i.e. $\phi = 1$), then the diffusion is described by Fick's law (equation (6.1)). If the scalar is *momentum* (i.e. $\phi = mv$), then the diffusion is due to viscous effects and described by Newton's law (equation (6.2)). If the scalar is *energy* (i.e. $\phi = mu$, where u is the specific internal energy)[12], then diffusion is a conduction process and described by Fourier's law (equation (6.3)).

One thing that the advection–diffusion form of the general transport equation is missing is an electric force that will move *charged* particles, which is especially relevant to gas discharges. As we will learn in chapter 7, the force of the electric field produces a different flux phenomenon called **drift**, which needs to be considered in conjunction with diffusion and advection.

6.2.2 Kinetic perspective—Boltzmann transport equation

The backbone of understanding transport from a kinetic perspective is the Boltzmann transport equation or BTE. Like the scalar transport equation, the BTE represents the general transport of a quantity. However, whereas the scalar transport equation considered the transport of the average quantity (ϕ), the BTE considers the transport of the energy distribution of the fluid particles $f(\varepsilon)$. More accurately, the scalar of interest is the velocity distribution $f(\vec{v})$, which can be related directly to the energy distribution (see section 2.2). This distribution can then be related to the scalar quantity ϕ. Ultimately, the BTE models the *non-equilibrium* behavior of the fluid particles, whereas the general scalar transport equation assumes that at every given point in space and time, the fluid is at equilibrium.

Consider figure 6.3, which shows a packet of particles moving through time. The packet initially has some distribution $f(\vec{r},\vec{v},t)$. That is, for any instant in time t, there exists a probability density function that describes the velocity distribution of

[12] What is meant by 'specific'? A specific quantity is a scalar per unit mass. For example, we can have internal energy per unit mass, entropy per unit mass, etc. Although we often think about physical properties, this concept can be in fact be non-intuitively applied to other common engineering quantities. For example, the specific form of momentum is velocity (momentum per unit mass).

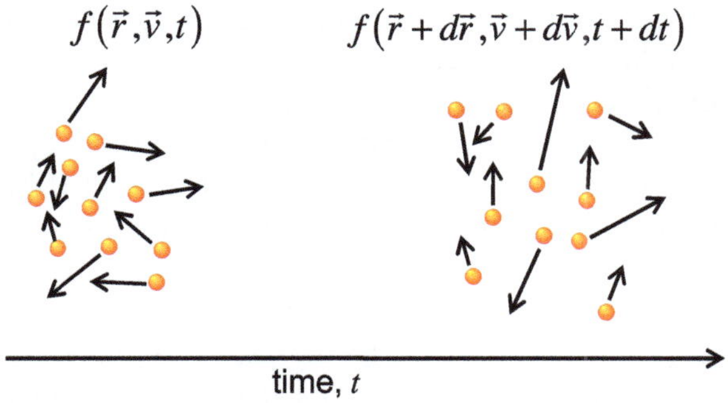

Figure 6.3. Illustration of the distribution of particles changing in time.

particles $f(\vec{v})$ at a specific position $\vec{r}$. However, as this packet of particles moves through time and space, the distribution at every point in time and space will evolve as particles undergo interactions with each other and their surroundings to obtain $f(\vec{r} + d\vec{r},\ \vec{v} + d\vec{v},\ t + dt)$. This distribution function f should obey a conservation equation in time, space, *and* velocity space. We will derive this in one dimension for simplicity and then generalize.

To a rough degree, we can say that the change in the distribution function f in time is

$$\frac{f(\vec{r} + d\vec{r},\ \vec{v} + d\vec{v},\ t + dt) - f(\vec{r},\ \vec{v},\ t)}{dt}. \tag{6.14}$$

If we consider each of the terms as applied to individual variables, we can see that $d\vec{r} = \vec{v}dt$ and $d\vec{v} = \vec{a}dt$, where $\vec{a}$ is the particle acceleration. We can therefore rewrite equation (6.14) as

$$\frac{f(\vec{r} + \vec{v}dt,\ \vec{v} + \vec{a}dt,\ t + dt) - f(\vec{r},\ \vec{v},\ t)}{dt}. \tag{6.15}$$

With some manipulation [3] it can be shown that this can be equated to

$$\frac{\partial f}{\partial t} + \vec{v}\cdot\nabla_{\vec{r}} f + \vec{a}\cdot\nabla_{\vec{v}} f = 0. \tag{6.16}$$

Here $\nabla_{\vec{r}}$ is the gradient in *physical space*, whereas $\nabla_{\vec{v}}$ is the gradient in *velocity space*[13].

A few notes on equation (6.17). This is called the **collisionless** BTE or *Vlasov equation*[14], because it only accounts for the transport of the distribution. Collisions

[13] That is, in Cartesian coordinates, $\nabla_{\vec{r}} = \frac{\partial}{\partial x}\hat{i} + \frac{\partial}{\partial y}\hat{j} + \frac{\partial}{\partial z}\hat{k}$ and $\nabla_{\vec{v}} = \frac{\partial}{\partial v_x}\hat{i} + \frac{\partial}{\partial v_y}\hat{j} + \frac{\partial}{\partial v_z}\hat{k}$, respectively.

[14] Named after Anatoly Alexandrovich Vlasov, a prominent Russian physicist. Technically, Vlasov adapted this equation specifically for plasmas by converting the generic acceleration term into its force equivalents using the Coulomb and Lorentz forces—sometimes called the Vlasov–Maxwell equation. He formulated this equation to study the effects of *long-range* interactions due to Coulombic forces between charged particles rather than the one-to-one collisions that we usually consider.

can also affect the distribution and will add more terms. Another thing to note is that f is a scalar, so this is directly analogous to the scalar transport equation except now there are seven dimensions—time, three in physical space, and three in velocity space. Finally, the acceleration term $\vec{a}$ can be correlated to an applied force —such as an electric or magnetic field—that will affect the distribution. Using the relationship $\vec{a} = \vec{F}/m$, equation (6.16) can also be written as

$$\frac{\partial f}{\partial t} + \vec{v}\cdot\nabla_{\vec{r}}f + \frac{\vec{F}}{m}\cdot\nabla_{\vec{v}}f = 0. \tag{6.17}$$

If one considers a collision between two particles, it will change both of their velocities $\vec{v}$ such that it is clear that the velocity distribution will also change. That is, the collisions act as a source/sink function to change the distribution. We can add this term as a change of the distribution in time to the right-hand side to obtain the general BTE,

$$\frac{\partial f}{\partial t} + \vec{v}\cdot\nabla_{\vec{r}}f + \frac{\vec{F}}{m}\cdot\nabla_{\vec{v}}f = \left.\frac{\partial f}{\partial t}\right|_c. \tag{6.18}$$

There are different ways to represent the collision term $\left.\frac{\partial f}{\partial t}\right|_c$ based on what are called integral formulations, and appendix B in [4] has a detailed discussion of this collision integral. In general, the BTE must be invoked when studying discharges that have non-equilibrium, that is, non-Maxwellian, particle distributions. Electrons in non-equilibrium, low-temperature gas discharges fit into this category.

6.2.3 From the Boltzmann transport equation to macroscopic properties

Once we have solved the BTE and know $f(\vec{v})$ for every given position and time $(\vec{r},t)$, we can extract macroscopic properties for the collection of particles for every given position and time $(\vec{r},t)$. To convert $f(\vec{r},\vec{v},t)$ to real properties, we multiply the property of interest by the distribution function and integrate over velocity space. For example, we obtain for the particle density simply by multiplying the distribution by 1 and integrating, or

$$N(\vec{r}, t) = \int f(\vec{r}, \vec{v}, t)d^3v. \tag{6.19}$$

Similarly, the particle flux is determined by multiplying the distribution by the velocity

$$\vec{J_N}(\vec{r}, t) = \int \vec{v}f(\vec{r}, \vec{v}, t)d^3v. \tag{6.20}$$

Other macroscopic properties follow the same principle, such as the average particle speed at $(\vec{r},t)$

$$v(\vec{r}, t) = \frac{\int \vec{v}f(\vec{r}, \vec{v}, t)d^3v}{\int f(\vec{r}, \vec{v}, t)d^3v}, \tag{6.21}$$

and the energy density (kinetic energy per unit volume)

$$w(\vec{r}, t) = \frac{1}{2}m \int v^2 f(\vec{r}, \vec{v}, t) d^3v, \tag{6.22}$$

which can be multiplied by the mean particle velocity to determine the energy flux[15].

We can use the same approach to actually derive the scalar transport equations from the BTE itself[16]. We do this by taking *moments* of the BTE, which means we multiply the BTE by a scalar of interest ϕ and integrate over all velocity space

$$\phi \int \left[\frac{\partial f}{\partial t} + \vec{v} \cdot \nabla_{\vec{r}} f + \frac{\vec{F}}{m} \cdot \nabla_{\vec{v}} f = \frac{\partial f}{\partial t}\bigg|_c \right] d^3v. \tag{6.23}$$

The detailed mathematics of this can be cumbersome [1] and are thus left to the reader.

6.3 Summary

The purpose of this chapter was to give an overview of general transport concepts and equations that are applicable to any system and will be invoked in this text for the study of discharges. There were a few primary points that were discussed:

- The flux of a scalar property can be described by kinetic concepts, and is qualitatively the flux of particles each carrying that scalar property.
- At the continuum scale, a general scalar transport equation can be used to describe the *conservation* of any scalar property by balancing the accumulation, flux, and generation/consumption of that property in a control volume.
- The Boltzmann transport equation (BTE) describes the evolution of the velocity distribution $f(\vec{v})$ of a collection of particles in time and space; the velocity distribution can further be related to the energy distribution $f(\varepsilon)$.
- The velocity distribution can be related to macroscopic properties by integrating the distribution over velocity space, and similarly the Boltzmann transport equation can be related to the general scalar transport equation by taking the moment of the Boltzmann transport equation.

It is important to recognize that the BTE and general scalar transport equation are intimately related. The BTE is always accurate in so much that there is a collection of particles that can be described by a distribution function. The general scalar transport equation is only appropriate when at any given position and time $(\vec{r}, t)$, the particles are in local equilibrium and can be described by their average properties (average velocity, average kinetic energy, etc). Taking the moment of the BTE is, in effect, averaging over the distribution $f(\vec{v})$. The general scalar transport equation is thus only appropriate when the system can be represented as a continuum, which occurs

[15] The energy density is in J m^{-3}, while the energy flux is in W m^{-2}.

[16] In fact, it can be shown that the constitutive laws shown in table 6.1 are special solutions of the BTE under the assumption of local thermodynamic equilibrium [1].

when there is a sufficient number of collisions such that the particles come to local equilibrium. We generally characterize this by the Knudsen number (Kn), defined as

$$Kn = \frac{\bar{l}}{d}, \tag{6.24}$$

where $\bar{l}$ is the mean free path and d is the characteristic length of the system (such as the distance between electrodes in a discharge). If $Kn \ll 1$, then there are sufficient collisions for the general scalar transport equation to be appropriate. However, if $Kn \sim 1$, implying that particles only experience on-average one single collision over a distance d, then the BTE must be used. And if $Kn \gg 1$, then the collisionless BTE is appropriate.

References

[1] Chen G 2005 *Nanoscale Energy Transport and Conversion: A Parallel Treatment of Electrons, Molecules, Phonons, and Photons* (New York: Oxford University Press)

[2] Deen W M 1998 *Analysis of Transport Phenomena* (New York: Oxford University Press)

[3] Fridman A and Kennedy L A 2011 *Plasma Physics and Engineering* 2nd edn (Boca Raton, FL: CRC Press)

[4] Lieberman M A and Lichtenberg A J 2005 *Principles of Plasma Discharges and Materials Processing* 2nd edn (New York: Wiley-Interscience)

Chapter 7

Transport equations for gas discharges

With the basic understanding of general transport concepts in chapter 6, we can apply the general transport equations specifically to gas discharges to determine their governing equations. In this chapter, we will approach this from the continuum-level, developing equations for ions and electrons that take the same form as the general scalar transport equation (6.12). We will also discuss where these equations are limited and how kinetic concepts, including the Boltzmann transport equation, come into play[1].

7.1 Equations for the charged particles

For our charged particles, positive and negative ions and electrons, there will be a set of equations that describe the conservation of particles/charge, momentum, and energy[2]. In the following sections, we will focus on those equations that describe positive ions (subscript *i*) and electrons (subscript *e*). We will neglect negative ions, though extending these general concepts to negative ions is straightforward.

7.1.1 Particle/charge conservation

If we start from the general transport equation, equation (6.12), we can write a general transport expression for ion conservation by setting $\phi = 1$ to obtain the conservation equation

$$\frac{\partial(N_i)}{\partial t} + \vec{\nabla}\cdot(\vec{J}_i) = S_{N_i}. \tag{7.1}$$

[1] A good resource that discusses the governing transport equations for low-temperature, non-equilibrium gas discharges is [2].

[2] This is directly analogous to the conservations of mass, momentum, and energy that form the governing equations in fluid dynamics.

doi:10.1088/978-0-7503-3991-9ch7

Note that we have left in a generalized flux $\vec{J}_i$. An alternative way to write the flux term is to relate it to the velocity of the ions, $\vec{J}_i = N_i\vec{v}_i$, such that equation (7.1) becomes

$$\frac{\partial(N_i)}{\partial t} + \vec{\nabla}\cdot(N_i\vec{v}_i) = S_{N_i}. \tag{7.2}$$

We can write an equivalent expression for our electrons

$$\frac{\partial(N_e)}{\partial t} + \vec{\nabla}\cdot(N_e\vec{v}_e) = S_{N_e}. \tag{7.3}$$

When we discussed the flux of particles in terms of a general scalar, we described it by advection and diffusion. However, in a discharge the charged particles are subject to external forces such as electric fields so that we must determine $\vec{v}_i$ and $\vec{v}_e$ explicitly using a momentum equation for each, which will be shown in the following section.

The *source* term in both these expressions is extremely important, that is $S_{N_i} \neq 0$ and $S_{N_e} \neq 0$. This source of ions and electrons arises from the volumetric ionization reactions that produce ions and electrons in the discharge, as discussed in chapter 4. For the time being, let us consider only the dominant electron-impact ionization process. The reaction can be written as

$$e^- + M \rightarrow 2e^- + M^+ \text{ iff } \varepsilon_e \geqslant E_{\text{ion}}.$$

The rate of generation (source term) can be written in terms of the reaction rate constant k_{ion} and the local concentration of the two reactants, in this case the gas N_g and electrons N_e, as

$$S_{N_i} = S_{N_e} = k_{\text{ion}}N_gN_e.$$

Note that the source term for both ions and electrons is *equivalent* because for each ion that is created, an electron is also created—they only come in pairs! However, the source term only depends on the electron number density N_e and not the ion density N_i. That is, the number of electrons already present in the discharge affects the reaction rate but the number of ions does not[3]. We can therefore write the two conservation equations as

$$\frac{\partial(N_i)}{\partial t} + \vec{\nabla}\cdot(N_i\vec{v}_i) = k_{\text{ion}}N_gN_e; \tag{7.4a}$$

$$\frac{\partial(N_e)}{\partial t} + \vec{\nabla}\cdot(N_e\vec{v}_e) = k_{\text{ion}}N_gN_e. \tag{7.4b}$$

However, as shown in equation (3.28), the reaction rate constant and reaction frequency are directly related by $\nu_j = N_gk_j$, and thus we can also write the conservation equations as

$$\frac{\partial(N_i)}{\partial t} + \vec{\nabla}\cdot(N_i\vec{v}_i) = \nu_{\text{ion}}N_e; \tag{7.5a}$$

[3] At least not directly in an electron-impact reaction.

$$\frac{\partial(N_e)}{\partial t} + \vec{\nabla}\cdot(N_e\vec{v_e}) = \nu_{\text{ion}}N_e. \tag{7.5b}$$

With suitable boundary conditions and knowledge of $\vec{v_i}/\vec{v_e}$ and ν_{ion}, these two equations can be solved for N_i and N_e. Note that equations (7.5a) and (7.5b) are explicitly coupled because N_e appears in both equations. Secondary emission is also a source of electrons but this only occurs at the electrodes of the discharge and thus can be treated as a *boundary condition*. Other gaseous sources of ions and electrons can simply be included as additional source terms on the right-hand side of the equations.

These equations can be converted into charge conservation by multiplying by Z_iq for the ions and $-q$ for the electrons. Here, Z_i is the charge per ion so that if an ion is singly charged, as in most cases, then $Z_i = 1$, but if the ion is doubly charged then $Z_i = 2$. Since electrons only carry a single unit charge by definition, $Z_e = -1$, always. Therefore, equations (7.5a) and (7.5b) can be transformed into equations for n_i and n_e as[4]

$$\frac{\partial(n_i)}{\partial t} + \vec{\nabla}\cdot(n_i\vec{v_i}) = -\nu_{\text{ion}}n_e; \tag{7.6a}$$

$$\frac{\partial(n_e)}{\partial t} + \vec{\nabla}\cdot(n_e\vec{v_e}) = \nu_{\text{ion}}n_e. \tag{7.6b}$$

Looking at this more closely, we realize that the flux of charge concentrations $n\vec{v}$ has units of A m^{-2}. That is, the fluxes in equations (7.6a) and (7.6b) are the *current densities* of ions and electrons, j_i and j_e, respectively. If we assume steady state, we can write these equivalently as equations for current density that can be solved directly for j_i and j_e. Recall from section 4.1 that ionization frequency can be related to Townsend's first ionization coefficient by the mean speed of the electrons or $\nu_{\text{ion}} = \bar{v_e}\alpha$. Therefore, equations (7.6a) and (7.6b) become

$$\vec{\nabla}\cdot(j_i) = -\alpha j_e; \tag{7.7a}$$

$$\vec{\nabla}\cdot(j_e) = \alpha j_e. \tag{7.7b}$$

Notice that in (7.7a), the source term became negative because electrons have a negative charge $-q$. Thus, we can see that these two equations imply that at steady state, $\vec{\nabla}\cdot(j_i) = -\vec{\nabla}\cdot(j_e)$. That is, the current densitites of ions and electrons are equal and opposite because they flow in opposite directions.

7.1.2 Momentum conservation

As noted above, we need an equation for the velocity of the ions and electrons, or more precisely, an equation that describes the *momentum* ($m\vec{v}$) of each. This conservation equation will again balance accumulation with flux and source terms,

[4] The charge densities n_i and n_e have units of C m^{-3}.

where the scalar in this case is a component of the momentum vector. The momentum equation for ions is

$$\frac{\partial(m_i N_i \vec{v}_i)}{\partial t} + \vec{\nabla}\cdot((m_i N_i \vec{v}_i)\vec{v}_i) = Z_i q N_i \vec{E} - \vec{\nabla} p_i + S_{\text{mom}}. \tag{7.8}$$

This yields a set of equations for each component of the mean particle velocity $\vec{v}_i(\vec{r}, t)$. On the left-hand side we see the familiar accumulation term and an advection flux term. On the right-hand side, instead of the familiar diffusion flux term, we see a **drift** flux term that relates the gain in momentum due to the force of the electric field $\vec{E}$. The pressure gradient is also a force that can induce momentum and is typically considered a source of momentum rather than a flux. Often, the pressure is described simply by using the ideal gas law so that it is related to the number density and average temperature of the particles

$$p = N k_b T. \tag{7.9}$$

That is, we invoke an equation of state[5]. However, much like how $\vec{v}_i$ appears in the continuity equation, now T_i will appear in the momentum equation and another equation is required to determine T_i—the energy equation.

Again, we can treat the source term as a 'frequency' of collisions because these 'create' momentum (transfer). However, whereas the source term in (7.5a) relies on frequency of *electron* collisions to create ions, this source term will rely on the frequency of *ion* collisions with other particles, as this makes them *lose* their momentum. We can thus define this source term as

$$S_{\text{mom}} = -\nu_{\text{mom},i}(m_i N_i \vec{v}_i).$$

The momentum equations for $\vec{v}_i$ and $\vec{v}_e$ thus can be written as

$$\frac{\partial(m_i N_i \vec{v}_i)}{\partial t} + \vec{\nabla}((m_i N_i \vec{v}_i)\vec{v}_i) = Z_i q N_i \vec{E} - \vec{\nabla}(N_i k_b T_i) - \nu_{\text{mom},i}(m_i N_i \vec{v}_i); \tag{7.10a}$$

$$\frac{\partial(m_e N_e \vec{v}_e)}{\partial t} + \vec{\nabla}((m_e N_e \vec{v}_e)\vec{v}_e) = -q N_e \vec{E} - \vec{\nabla}(N_e k_b T_e) - \nu_{\text{mom},e}(m_e N_e \vec{v}_e). \tag{7.10b}$$

Equations are now needed for T_i and T_e and also $\vec{E}$ to create a closed system of equations.

7.1.3 Energy conservation

As noted above, we need an equation for the temperature of the ions and electrons, or more precisely, an equation that describes the *energy* (ε) of each species. Such a

[5] We can invoke an ideal gas under the *isothermal* assumption, that is, the temperature is always equilibrated locally. Most discharge analyses can make this approximation because the processes are relatively 'slow'. However, if the phenomena of interest is 'fast' such as in ionization waves, then typically this local equilibrium assumption is not valid, and a different equation of state is needed. Often it is assumed that in fast phenomena, the gas does not exchange energy with its surroundings and an *adiabatic* assumption is made such that the equation of state is $p = CN^{\gamma}$, where C is a constant and γ is the ratio of specific heats. In this case, temperature is not introduced, and an energy equation is not needed to close the system of equations.

conservation equation will again balance accumulation with flux and source terms. The energy equation for ions is

$$\frac{\partial(N_i\varepsilon_i)}{\partial t} + \vec{\nabla}\cdot(N_i\varepsilon_i\vec{v}_i) = \vec{\nabla}\cdot(\kappa_i\vec{\nabla}T_i) + (qN_i\vec{v}_i)\cdot\vec{E} + S_{\text{energy}}. \tag{7.11}$$

This yields an equation for the mean ion energy ε_i. On the left-hand side we see the familiar accumulation term and an advection flux term. On the right-hand side, we see the familiar thermal conduction (diffusion) flux term as well as a source term related to the electric field. This is a **Joule** (or **ohmic**) **heating** term due to the kinetic energy added to the ions by the electric field.

The last source term is energy exchange due to collisions and depends on the type of particles that are being considered. If we are considering large ions colliding with neutral particles and/or other ions, the collisions can usually be treated as elastic collisions where there is kinetic energy loss, and the source term is

$S_{\text{energy,elastic}} = -\nu_{\text{elastic},i}N_i\varepsilon_i$.

While electrons can also lose kinetic energy via elastic collisions with neutral particles, they will also lose energy through *inelastic* collisions, i.e. excitation and ionization reactions. In this case, the electrons lose the threshold energy E_j of reaction j and the rate can be written as

$S_{\text{energy,inelastic}} = -k_jN_gN_eE_j$.

Finally, if we invoke a Maxwellian assumption for energy[6], we can let $\varepsilon = \frac{3}{2}k_bT$. We can thus write the energy equation for ions as

$$\frac{\partial\left(\frac{3}{2}k_bT_iN_i\right)}{\partial t} + \vec{\nabla}\cdot\left(\frac{3}{2}k_bT_iN_i\vec{v}_i\right) = \vec{\nabla}\cdot(\kappa_i\vec{\nabla}T_i) + (Z_iqN_i\vec{v}_i)\cdot\vec{E} - \nu_{\text{elastic},i}\left(\frac{3}{2}k_bT_iN_i\right), \tag{7.12}$$

and for electrons as

$$\begin{aligned}\frac{\partial\left(\frac{3}{2}k_bT_eN_e\right)}{\partial t} + \vec{\nabla}\cdot\left(\frac{3}{2}k_bT_eN_e\vec{v}_e\right) &= \vec{\nabla}\cdot(\kappa_e\vec{\nabla}T_e) + (-qN_e\vec{v}_e)\cdot\vec{E} \\ &\quad - \nu_{\text{elastic},e}\left(\frac{3}{2}k_bT_eN_e\right) - \sum_j k_jN_gN_eE_j.\end{aligned} \tag{7.13}$$

Again, the energy equations are clearly coupled to both the particle conservation and momentum equations.

7.1.4 Electric field

The only remaining unknown in this set of equations is the electric field $\vec{E}$ which acts like a force in the momentum equations and a source of kinetic energy in the energy equations. As discussed in the chapter 1 this is governed by Poisson's law

[6] That is, we assume that the energy at every position $\varepsilon(\vec{r},t)$ is the average of the local particle distribution.

$$\nabla^2\Phi = -\frac{(Z_iqN_i - qN_e)}{\epsilon_0}, \tag{7.14}$$

where the electric field is related to the potential by $\vec{E} = -\vec{\nabla}\Phi$. Combining these two expressions, we can write equation (7.14) explicitly in terms of the electric field,

$$\vec{\nabla}\cdot\vec{E} = \frac{(Z_iqN_i - qN_e)}{\epsilon_0}. \tag{7.15}$$

We can now state that we have a complete set of continuum equations for ions, electrons, and the electric field consisting of particle (equations (7.4a) and (7.4b)), momentum (equations (7.10a) and (7.10b)), and energy (equations (7.12) and (7.13)) conservation, and Poisson's equation. These equations describe the transport in a discharge from an equilibrium perspective, and can be solved for the evolution of $N(\vec{r},t)$, $\vec{v}(\vec{r},t)$, and $T(\vec{r},t)$ in space and time. Similar equations can be written for negative ions or any of the excited species produced by electron-impact collisions as well as the neutral (ground state) gas particles.

Mathematically, these equations are coupled and highly nonlinear, making them challenging to solve analytically, and thus difficult to describe discharge behavior with mathematical solutions. (In fact, if we only consider ions and electrons and only one dimension, there are still seven equations, whereas in three dimensions there are 11 equations that would need to be solved simultaneously.) Furthermore, these equations all depend on either reaction rates or collision frequencies, which themselves depend on the energy distribution of the particles. Thus, to be even more complete, we need expressions for all the k and ν terms, which will either come from empirical expressions or from solutions to the BTE. We therefore seek assumptions that lead to simplifications of the equations that allow us to gain insight into basic discharge behavior.

7.2 The drift–diffusion approximation

The most common simplification is the drift–diffusion approximation, which essentially combines the three conservation equations into a single particle conservation equation, thus reducing the equation set considerably. What we seek is a simple solution to the momentum equation, such that we have a simple expression for the particle flux, $\vec{J} = N\vec{v}$, which we can plug into the particle conservation equation.

Consider the steady state momentum equation. If we neglect inertia (as is often the case for low-mass electrons in particular), assume a constant (that is, not energy-dependent) collision frequency, and assume a *uniform* temperature for our particles, we can reduce the momentum equation for ions to

$$Z_iqN_i\vec{E} - k_bT_i\vec{\nabla}(N_i) - \nu_{\text{mom},\,i}m_iN_i\vec{v}_i = 0. \tag{7.16}$$

What are the implications of these assumptions? By assuming a uniform temperature distribution, we have essentially removed the need to solve the energy equation. (That is, setting T_i and T_e to be constant and uniform is a trivial solution to the energy equation.) Thus, we have already removed one of the conservation equations.

Similarly, by assuming a constant collision frequency, we have removed the need to determine the energy distribution of our particles—that is, we have assumed some average temperature. For ions, for example, we might assume that they have the same temperature as the neutral gas.

In this form, equation (7.16) can be solved directly for the velocity $\vec{v}_i$,

$$\vec{v}_i = \frac{Z_i q\vec{E}}{\nu_{\mathrm{mom},i} m_i} - \frac{k_b T_i \vec{\nabla} N_i}{\nu_{\mathrm{mom},i} m_i N_i}. \tag{7.17}$$

If we multiply by N_i this simply becomes the flux $\vec{J}_i$ or

$$\vec{J}_i = N_i \vec{v}_i = \frac{Z_i q\vec{E}}{\nu_{\mathrm{mom},i} m_i} N_i - \frac{k_b T_i}{\nu_{\mathrm{mom},i} m_i} \vec{\nabla} N_i. \tag{7.18}$$

The flux, therefore, is divided into two terms. By analogy with the advection and diffusion components of the flux see equation (6.13), we see that the two terms on the right-hand side are the drift and diffusion components of the flux, where the drift is defined as

$$\textbf{Drift}: \vec{J}_{\mathrm{drift}} = \frac{Z_i q\vec{E}}{\nu_{\mathrm{mom},i} m_i} N_i = \mu_i \vec{E} N_i$$

and the diffusion is

$$\textbf{Diffusion}: \vec{J}_{\mathrm{drift}} = -\frac{k_b T_i}{\nu_{\mathrm{mom},i} m_i} \vec{\nabla} N_i = -D_i \vec{\nabla} N_i.$$

The diffusion expression contains the familiar diffusivity D_i, and the drift expression contains a similar parameter called the **mobility**, μ_i. The mobility can be thought of as the ease with which a charged particle moves through a medium and can be related to the electrical conductivity of the discharge. From these expressions, we can see that the mobility and diffusivity are defined as

$$\mu_i = \frac{Z_i q}{\nu_{\mathrm{mom},i} m_i}; \tag{7.19a}$$

$$D_i = \frac{k_b T_i}{\nu_{\mathrm{mom},i} m_i}. \tag{7.19b}$$

If we relate the mobility and diffusivity by noting they both contain $\nu_{\mathrm{mom},i} m_i$ in the denominator, we arrive at what is known as Einstein's relation[7]:

$$\mu_i = \frac{Z_i q}{k_b T_i} D_i. \tag{7.20}$$

[7] Named after Albert Einstein, who determined the relationship during his studies on Brownian motion as part of his seminal papers in 1905. It is sometimes also referred to as the Einstein–Smoluchowski relation because Polish physicist Marian Smoluchowski independently derived the same relationship at about the same time.

A similar expression can also be used for electrons.

This is called the **drift–diffusion approximation**, and we can now rewrite the particle conservation equations as

$$\frac{\partial(N_i)}{\partial t} + \vec{\nabla}\cdot(\mu_i\vec{E}N_i) = \vec{\nabla}\cdot(D_i\vec{\nabla}N_i) + \nu_{\text{ion}}N_e; \tag{7.21a}$$

$$\frac{\partial(N_e)}{\partial t} + \vec{\nabla}\cdot(\mu_e\vec{E}N_e) = \vec{\nabla}\cdot(D_e\vec{\nabla}N_e) + \nu_{\text{ion}}N_e. \tag{7.21b}$$

This drift–diffusion approximation greatly simplifies the analysis of the discharge by wrapping the momentum and energy equations directly into the continuity equation. Thus, to describe ions and electrons, one only needs these two equations and an equation for the electric field (equation (7.15)).

The current density in the drift–diffusion approximation is defined as

$$\vec{j} = \mu\vec{E}n - D\vec{\nabla}n.$$

Often diffusion is negligible so this simplifies to

$$\vec{j} = \mu\vec{E}n.$$

From this expression, we see that the *average* velocity of the charged particles is $\vec{v} = \mu\vec{E}$. We define this as the *drift* velocity or

$$\vec{v}_{\text{drift}} = \mu\vec{E}. \tag{7.22}$$

It should be clear, then, that the units of mobility are $\text{m}^2\ \text{V}^{-1}\ \text{s}^{-1}$.

7.3 Summary

The purpose of this chapter was to derive the transport equations that describe a discharge. A few primary points were discussed:

- The conservation of particles/charge, momentum, and energy can be derived for all the charged species along with Poisson's equation for the electric field to arrive at a complete set of continuum equations that describe the conservation of species and properties in a discharge.
- A drift–diffusion approximation can be applied to reduce the conservation equations to only particle conservation, where the ion and electron fluxes are described by electric drift and diffusion.
- We can estimate the velocity of ions and electrons based on their drift alone, producing a drift velocity, under the assumption that diffusion is relatively small.

It is important to realize that this chapter only considered the transport equations at the continuum scale. However, the kinetic BTE is also very important, primarily because we can often *not assume* local thermal equilibrium for the electrons. Solving the BTE for the electrons is inherently difficult due to the inherent complexity of the

BTE as a seven-dimensional integro-differential equation. One of the more common approaches is to invoke the **local field approximation** (LFA), which assumes that the distribution function for the electrons $f(\varepsilon)$ only depends on the local electric field and number density, and thus the reduced electric field E/N. A detailed discussion of the LFA, its validity, and the resulting solution BTE is beyond the scope of this text, but the reader is pointed to several good discussions and models, including [1–3].

References

[1] Hagelaar G J M and Pitchford L C 2005 Solving the Boltzmann equation to obtain electron transport coefficients and rate coefficients for fluid models *Plasma Sources Sci. Technol.* **14** 722–33

[2] Kushner M J 2009 Hybrid modelling of low temperature plasmas for fundamental investigations and equipment design *J. Phys. D: Appl. Phys.* **42** 194013

[3] Colonna G 2016 Boltzmann and Vlasov equations in plasma physics *Plasma Modeling: Methods and Applications* ed G Colonna and A D'Angola (Bristol: IOP Publishing) http://iopscience.iop.org/book/978-0-7503-1200-4/chapter/bk978-0-7503-1200-4ch1

Part III

Gas discharges and plasmas

IOP Publishing

Ionization and Ion Transport (Second Edition)
A primer for the study of gas discharges and plasmas
David B Go

Chapter 8

Discharge initiation and plasma formation

With an elementary understanding of the basic underlying processes in a gas discharge, including how the charged particles are *formed* (ionization) and *move* (ion transport), we can now turn our attention to gas discharges themselves. The first question at hand is: how is a discharge formed? Once that is answered, the next basic question is: what are some of the basic plasma properties? This part of the book is organized into three chapters. Chapter 8 deals with plasma initiation or how gas discharges are formed. Chapter 9 then considers some of the basic structure of a plasma and chapter 10 describes the most common discharge of all, the dc glow discharge, how it is practically formed, and its basic properties.

As discussed in chapter 1, this book is concerned with gas discharges that are generated in laboratories. That is, artificial plasmas formed by the application of an electric or magnetic field[1]. Recall, a gas discharge is essentially an electrical flow (current) across an interstitial gap of gas, turning the insulating gas into a conductor. This does not require that the gas is in a plasma state, but at some point, under some conditions, the gas will transition into a plasma. What is the mechanism of this transition? What properties does it depend on? Can we predict this transition? In this chapter, we will overview discharge initiation leading to breakdown and the formation of a plasma. This will build directly upon the physics discussed in parts I and II, and in particular, the equations used to describe electron and ion behavior in chapter 7.

8.1 Townsend behavior below the breakdown threshold

The process of creating a discharge begins with the formation of an electron avalanche. That is a single free electron is accelerated by an electric field, collides

[1] There are of course other ways to generate discharges, even in laboratories, including using lasers or microwaves, or even more exotically, triboelectric charging or piezoelectric crystals.

doi:10.1088/978-0-7503-3991-9ch8

with a neutral molecule and creates a second electron. They are both then accelerated by the electric field and create two more electrons. This α-process results in an exponential growth in the number of electrons, as discussed in chapter 4 (equation (4.9)) and shown in figure 4.2. Is that enough to create a discharge? What happens when all the electrons finally reach the anode across the electrode gap from the cathode? If there is nothing else going on, the answer is 'no'—it is not enough to create a discharge. However, as we learned in chapter 5, there is something else going on. The ions that are produced in the gap move toward the cathode and can produce more electrons from the cathode via secondary emission, the γ-process (figure 5.1). There is a continual process of electrons being produced by secondary emission at the cathode, multiplying via ionization in the electrode gap, and then leaving through the anode. We can treat this mathematically as follows.

Recall from section 7.1 that the discharge can be described by a set of equations for the ion and electron densities that includes the storage, transport, and generation of electrons and ions. If we assume steady state and use the form of the equations with current density, equations (7.7a) and (7.7b) can be written in one-dimensional (1D) form as

$$\frac{dj_i}{dx} = -\alpha j_e; \tag{8.1a}$$

$$\frac{dj_e}{dx} = -\frac{dj_i}{dx}. \tag{8.1b}$$

Also recall from chapter 4 that α is a function of the reduced electric field E/p (equation (4.15))

$$\alpha = Ap \exp\left[-\frac{Bp}{E}\right]. \tag{8.2}$$

and thus we also need to include an equation for the potential and electric fields from Poisson's equation (7.15). In 1D we can write this as

$$\frac{d^2\Phi(x)}{dx^2} = -\frac{q}{\varepsilon_0}(N_i(x) - N_e(x)); \tag{8.3a}$$

$$E_x(x) = -\frac{d\Phi(x)}{dx}, \tag{8.3b}$$

where the electron and ion densities are related to the current density by $j(x) = \mu E(x)qN(x)$. Equations (8.1a)–(8.3b) form a set of first order, non-linear coupled ordinary differential equations, where the non-linearity is introduced by the form of Townsend's first ionization coefficient α and the coupling to Poisson's equation.

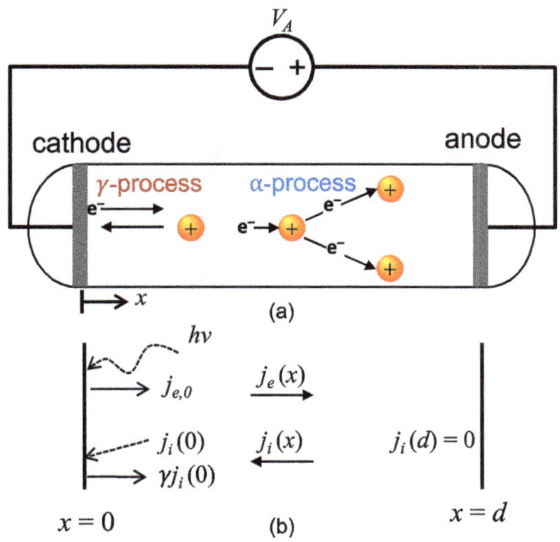

Figure 8.1. (a) Schematic of a basic dc discharge device consisting of a vacuum chamber held at pressure p, two electrodes separated by a distance d, and a dc power supply applying an applied voltage V_A between the two electrodes. (b) Illustration of boundary conditions outlined in equations (8.4) and (8.5).

Because both transport equations as well as the expression for the electric field are first order, we only require one boundary condition for each. If we take the simple dc system outlined in figure 8.1, we can analyze the basic physical behaviors relatively easily. First, for the ions we can assume that all ions are produced by ionization *within* the electrode gap and that they then move under the influence of the electric field to the cathode. Thus, there are no ions coming into or out of the system through the anode ($x = d$) and

$$j_i(x = d) = 0. \tag{8.4}$$

For the electrons, as discussed above, they can be produced at the cathode by secondary emission. Further, as we learned in chapter 5, the number of emitted electrons depends on the number of incident ions[2]. We can thus write that the total *flux* of electrons leaving the cathode depends on the total *flux* of ions arriving at the cathode, or

$$j_e(x = 0) = \gamma j_i(x = 0) + j_{e,0}. \tag{8.5}$$

The third term in this expression, $j_{e,0}$, describes a small background cathode emission current density[3]. It is largely negligible, but convenient mathematically.

[2] Which are related by the secondary emission coefficient γ (equation (5.22)).

[3] This background current was produced experimentally in Townsend's original experiments by photoemission at the cathode.

The electric field depends on the potential between each of the electrodes, which we will call the applied voltage (V_A),[4] and the boundary condition can be written as

$$\int_0^d -E(x)dx = \Phi(d) - \Phi(0) = V_A. \tag{8.6}$$

If we assume that conditions are such that there is minimal ionization, then the number density terms in equation (8.3a) are negligible, and the magnitude of the electric field equivalently becomes

$$| E | = \frac{V_A}{d}. \tag{8.7}$$

Under this circumstance, the α coefficient becomes a constant

$$\alpha = Ap \exp\left[-\frac{Bpd}{V_A}\right]. \tag{8.8}$$

Note that in the numerator of the exponential argument is the product of pressure p and electrode gap distance d, which is often just called pd for convenience[5]. This will be important later.

We can solve equations (8.1a) and (8.1b) by integrating and applying the boundary conditions in equations (8.4) and (8.5) to obtain

$$j_e(x) = \frac{j_{e,0}}{1 - \gamma(e^{\alpha d} - 1)} e^{\alpha x}; \tag{8.9a}$$

$$j_i(x) = \frac{j_{e,0}}{1 - \gamma(e^{\alpha d} - 1)} [e^{\alpha d} - e^{\alpha x}]. \tag{8.9b}$$

These two equations describe the distribution of current density for electrons and ions throughout the interstitial gap for a given pressure, gap distance, and applied voltage. Not surprising, they both reflect the **electron avalanche** behavior discussed in chapter 4. Because they rely only on the Townsend mechanisms of the α-process and γ-process, this set of equations is sometimes referred to as a Townsend formulation describing Townsend behavior.

Notice how the Townsend expressions deviate from the electron avalanche expressions in chapter 4 by a very important pre-factor of $\frac{1}{1 - \gamma\left(e^{\alpha d} - 1\right)}$. Because α is a function of the applied voltage in equation (8.8), the pre-factor is *also* a function of the applied voltage. Further, if we look closely at the pre-factor, it depends on

[4] Technically this is the applied voltage *difference* or potential difference because it represents the difference in potentials at the cathode and anode. Practically, the cathode or anode is typically grounded at zero potential ($\Phi = 0$ V) and a negative (cathode) or positive (anode) potential is applied at the other electrode. It is often also called a 'bias'.

[5] Pronounced 'pee-dee'.

both α and γ, which means that the current densities and ion and electrons concentrations depend on *both* the type of gas (α) *and* the material of the cathode (γ). Finally, notice that if we incorporate the expression for α from equation (8.8), then the argument of the exponential term becomes

$$\alpha d = Apd\exp\left[-\frac{Bpd}{V_A}\right]. \tag{8.10}$$

That is, the ionization in the electrode gap depends on the product of **pressure times distance**, or *pd*.[6] Further, per equation (8.7), the applied electric field is $E_A = \frac{V_A}{d}$, and the argument of the exponential then becomes

$$\exp\left[-\frac{Bp}{E_A}\right].$$

The ratio of the electric field to the pressure, E/p, is known as the **reduced electric field**[7]. These represent two of the most important scaling parameters in gas discharges.

8.2 Pressure times distance and reduced electric field scaling

Pressure times distance, *pd*, reflects the collisionality of the system; that is, how frequently electrons experience collisions over the distance d between electrodes. Recall that the mean free path in equation (2.15) is inversely proportional to pressure. Thus *pd* roughly scales as the ratio of the electrode distance to mean free path ($pd \sim d/\bar{l}$). If *pd* is *large*, then over the distance d, an electron will experience many collisions. If *pd* is *small*, then over the distance d, an electron will experience very few collisions. Recall from chapter 2 that these collisions will not only be ionization events, but also elastic collisions and excitation collisions (as well as electron attachment, dissociation, etc). The key is that as long as *pd* is the same across two different discharge configurations, electrons in the discharge will experience the same collisional environment or the same average number of collisions between the two electrodes.

Whereas *pd* describes how many collisions occur an electron experiences under a given set of conditions, the reduced electric field, E/p, describes how much energy an electron gains between collisions. This becomes clear when the electric field is replaced by V/d and written as

$$\frac{E}{p} = \frac{V}{pd}.$$

If *pd* is the average number of collisions, then V is the average potential gained between each collision. Within the context of equation (8.10), this is treated as a

[6] Typically expressed in units of Torr cm.

[7] The units of reduced electric field are conventionally written as V Torr^{-1} cm^{-1} (volts per *pd*). However, an alternative formulation is E/N, where N is the gas density, which is directly proportional to pressure, and has units of cm^{-3}. In this formulation, the unit of reduced electric field is given a special name of *Townsend* (Td), where 1 Td = 10^{-17} V cm^2.

global parameter based on the applied voltage V_A and electrode gap d. But it can actually be treated as a local parameter, based on the local electric field and number density of the gas; many properties, including local ionization rates and transport properties depend on the local E/p.

Overall, these two scaling parameters can be used to describe two experiments with different operating conditions where the behavior of both systems will essentially be the same. For example, consider set-up A where the electrodes are 1 cm apart and the gas pressure is 10 Torr such that pd = 10 Torr cm. Now consider a set-up B where the electrodes are 1 mm apart at a pressure of 100 Torr, and again pd = 10 Torr cm. If *everything else is the same*, including the applied voltage as well as the type of gas or gas mixture, then αd in equation (8.10) will be equivalent. Therefore, the exact same number of electrons will be produced by ionization in both set-ups. Now to be clear, this is not universally true and only applies over a range of pd where no other physics are relevant. Typically, most common gas discharge devices operate roughly over a range from 1 to 500 Torr cm, and values below or above that range require additional considerations, which is particularly relevant during breakdown, as will be shown in the following.

8.3 Breakdown

Equations (8.9a) and (8.9b) describe the current density in a one-dimensional dc discharge and most importantly, as shown in equation (8.7), assuming no space charge. Conceptually, what these equations describe is what happens when some seed electrons at the cathode (denoted by $j_{e,0}$) are placed in a potential field. They are accelerated by the electric field and collide with the gas to produce ions (the α-process), creating exponentially more electrons until they reach the anode. They are then replaced by the ions moving back toward the cathode, where more electrons are released by secondary emission (the γ-process) and the whole sequence continues. That is in essence a gas discharge, where electricity is conducted across the gas.

Observe that the current densities are exponential functions of both α and d, and as noted above, therefore dependent on the applied voltage V_A. As V_A is increased, these current densities continue to grow exponentially. If we look closely at the plot of current densities in figure 8.2, we see that at as the applied voltage increases, at some value, the current densities begin to grow extremely quickly. This rapid growth is due to the pre-factor $\frac{1}{1-\gamma\left(e^{\alpha d}-1\right)}$; as the voltage increases, the denominator becomes small and the pre-factor grows. When the denominator goes to zero, the current becomes so amplified, that it shoots to infinity. Now this cannot happen in real life—an infinite current is not physically possible—and at that condition the actual physical circuit comes into play. The current will be limited either by the power supply providing the applied voltage or some other circuit elements. Furthermore, the basic assumptions we made to derive the equations no longer apply. We have created so much current by ionization and secondary emission that the concentration of ions and electrons will be large and space charge comes into play; equation (8.7) is no longer valid.

We call this transition **breakdown** and it is often referred to when the gas transitions into a plasma, and the voltage where this occurs is called the **breakdown**

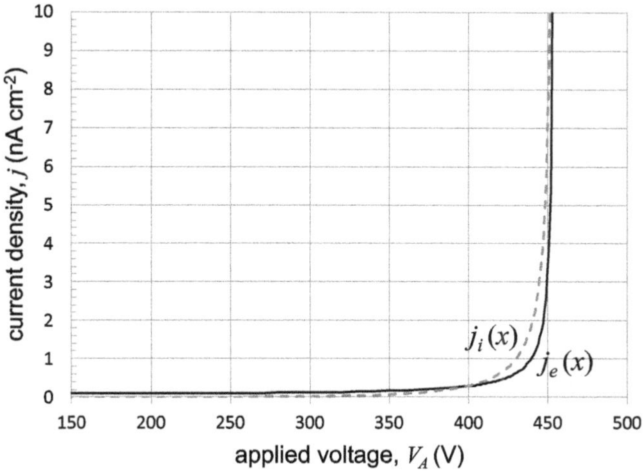

Figure 8.2. Plot of electron (j_e) and ion (j_i) current densities using equations (8.9a) and (8.9b), respectively, as a function of the applied voltage at a position of $x = d/2$. The data are for argon using the gas constants of $A = 11.5$ Torr^{-1} cm^{-1} and $B = 176$ V Torr^{-1} cm^{-1} [1] and the secondary emission coefficient $\gamma = 0.5$ assuming a background cathode emission current of $j_{e,0} = 0.1$ nA cm^{-2}. A pressure of $p = 10$ Torr and a gap distance of $d = 1$ cm were assumed for a $pd = 10$ Torr cm.

voltage[8]. Here we will describe the two simple cases of breakdown under a dc applied voltage between two flat plates. The first is *Townsend breakdown* which occurs under relatively modest *pd* values and the second is *streamer breakdown*, which occurs at large *pd* values. It goes without saying that breakdown can also occur under AC fields or by any other stimuli that accelerate electrons to create electron avalanches, including microwaves and lasers. These will not be discussed here but can be found in other texts and review articles.

8.3.1 Townsend breakdown

Townsend breakdown occurs when the two Townsend mechanisms—the α-process and the γ-process—are dominant and the conditions of the Townsend discharge are valid. We can thus analyze the critical voltage condition where the amplification leads to breakdown by analyzing the pre-factor itself. Amplification will hit a critical threshold when the denominator is equal to 0 or

$$1 - \gamma(e^{\alpha d} - 1) = 0. \tag{8.11}$$

Equation (8.11) therefore, is the **Townsend breakdown criterion**; when this is met the current becomes overamplified. But what does this mean physically? With some rearranging, we can rewrite equation (8.11) as

$$(e^{\alpha d} - 1) = 1/\gamma. \tag{8.12}$$

[8] Or the breakdown field in terms of the electric field.

We know from equation (4.9) in chapter 4 that the left-hand side of this equation is electron avalanche process for a single electron. That is if a single electron is released from the cathode, it will produce $(e^{\alpha d} - 1)$ ions. For the right-hand side, the know from equation (5.22) in chapter 5 that the secondary emission coefficient is the number of emitted electrons per incident ion. Therefore, the inverse of this, $1/\gamma$, the number of incident ions to produce one secondary emitted electron. We could thus describe equation (8.12) as follows:

ions created via ionization by a single electron = number of ions needed to emit a single electron via secondary emission.

That is, the breakdown criterion is when a single emitted electron produces enough ions to replace itself via secondary emission. Another way to describe this is that breakdown occurs when the discharge is **self-sustained**, the loss of electrons at the anode is completely balanced by the emission of electrons at the cathode.

While equation (8.11) describes the breakdown criterion, it does not actually provide the practical conditions (i.e. applied voltage), when breakdown occurs. However, if we plug equation (8.10) into (8.11) and use some clever mathematical manipulation[9], we can extract the applied voltage where breakdown occurs:

$$V_b = \frac{Bpd}{\ln(Apd) - \ln\left[\ln\left(\frac{1}{\gamma} + 1\right)\right]}. \tag{8.13}$$

Equation (8.13) is known as **Paschen's Law**[10], and has several important features. First, as expected, it depends on the gas type through the coefficients A and B in the expression for α and on the cathode material through γ. Second, it also depends on pd, the universal scaling parameter discussed previously. What this means is that for a given gas, the breakdown voltage not only depends on both the pressure *and* distance between electrodes, but their product. Thus, any two conditions that have matching pd, such as 1 cm at 10 Torr and 0.1 cm at 100 Torr, will have the same breakdown voltage.

Figure 8.3 shows a plot of equation (8.13) for argon, which is known as a **Paschen curve**; we notice a few particular features. First, the curve is non-monotonic; there is a left branch, a right branch, and a minimum. Notice that right branch, which is at larger pd values (1–100 Torr cm) is relatively linear. The slope of this branch is V_b/pd, which can be written equivalently as

$$\frac{E_b}{p} = \frac{V_b}{pd}, \tag{8.14}$$

[9] This is actually a great exercise to do on one's own time.

[10] Named after the German physicist Friedrich Paschen, who first discovered the breakdown behavior empirically in 1889 [2]. The mathematical formulation was not derived until Townsend did so in 1915 [3].

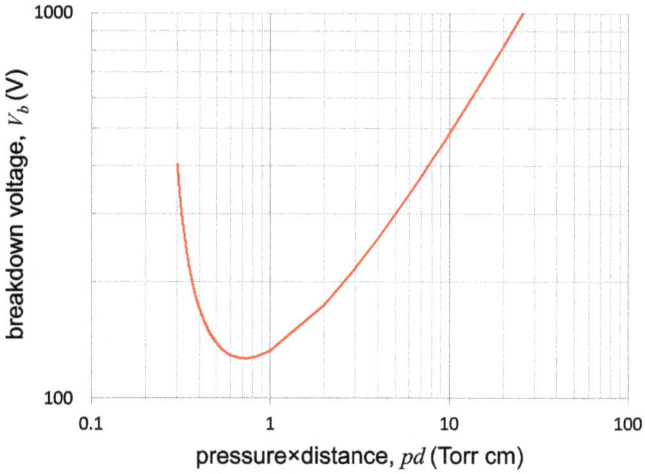

Figure 8.3. Paschen curve for argon using the gas constants of A = 11.5 Torr^{-1} cm^{-1} and B = 176 V Torr^{-1} cm^{-1} [1] and the secondary emission coefficient $\gamma = 0.5$.

and is the reduced electric field needed to achieve breakdown[11]. That is, as long as the pd conditions are relatively reasonable, breakdown always occurs at the same electric field at a given pressure. For example, in atmospheric pressure air, the breakdown field is roughly 30 kV cm^{-1}.

The left branch is also very peculiar, as it asymptotically increases as pd becomes very small. As discussed before, at small pd, an electron experiences very few collisions over the distance between electrodes. Therefore, there are far fewer chances for ionization and the opportunity to create enough ions to induce secondary emission. The applied potential must be much greater to ensure electrons are accelerated enough between each precious collision to increase the probability of ionization.

The minimum is a very curious part of the Paschen curve. This point, which is called Stoletov's point, represents the pd condition that requires the lowest amount of input energy to breakdown. The minimum can be determined by taking the derivative of equation (8.13) and setting it equal to zero, leading to a minimum applied potential of

$$(V_b)_{\min} = \frac{\bar{e}B}{A}\ln\left(\frac{1}{\gamma} + 1\right), \tag{8.15}$$

where $\bar{e} = \exp(1)$. This minimum occurs at a pd value of

$$(pd)_{\min} = \frac{\bar{e}}{A}\ln\left(\frac{1}{\gamma} + 1\right), \tag{8.16}$$

[11] Where E_b is the breakdown electric field.

and corresponds to a reduced electric field of

$$\left(\frac{E_b}{p}\right)_{\min} = B. \tag{8.17}$$

This last result is interesting in that the gas constant B in the expression for α is, by definition, the minimum reduced electric field where it is possible to breakdown a gas.

It should be noted that Paschen's law is not globally applicable; it is only valid for conditions where the dominant processes are electron avalanche formation by the α-process and secondary emission by the γ-process. However, in high vacuum ($pd < 10^{-3}$ Torr cm), the breakdown process cannot rely on these Townsend processes because there simply are not enough gas particles. Generally, vacuum breakdown occurs because of field emission (as discussed in chapter 5), where there is a very high electric field (often requiring an applied voltage of hundreds of kV). This is often accompanied by damage to the electrodes where material is physically evaporated away or desorbed gases provide additional gas-phase particles for ionization and other processes [4].

Another extremely small pd condition is when the electrode gap is very small, on the order of microns (~1–10 μm). Such a condition can occur not only in plasma devices—so-called microplasmas—but in other applications such as microelectro-mechanical systems (MEMS). At these dimensions, field emission is also necessary to produce electrons from the cathode, but because of the very small gap, only requires a modest applied voltage. The left branch in the Paschen curve in figure 8.3 actually continues to decrease, rather than increase, nearly linearly toward zero, forming what has been often called the 'modified Paschen curve' [5].

Paschen's law can also fail for very large electrode gaps when pd is large and 1D assumptions are no longer valid [6]. At high pressures, pd can also become large, and there is enough ionization in the electrode gap that other factors affect electron avalanches and streamers form, which is covered in the next section.

8.3.2 Streamer breakdown

When there is excessive ionization, there can be a buildup of *space charge* (positive or negative charges). As we know from Poisson's equation, equation (8.3a), space charge will modify the potential distribution and therefore, the electric field. The modified electric field influences the behavior of electrons and ions in the gap, and individual electron avalanches merge to form highly ionized channels. That is, breakdown is highly localized to one part of the gap (and one spot on the electrodes), rather than uniform across the electrode gap. Generally, two types of streamer mechanisms exist—anode directed streamers that start at the cathode, also called negative streamers, and cathode directed streamers that start at the anode, also called positive streamers.

It is easiest to qualitatively describe an anode directed streamer by considering figure 8.4. In figure 8.4(a), the typical electron avalanche is depicted, where a single electron starting at the cathode moves toward the anode, creating more electrons

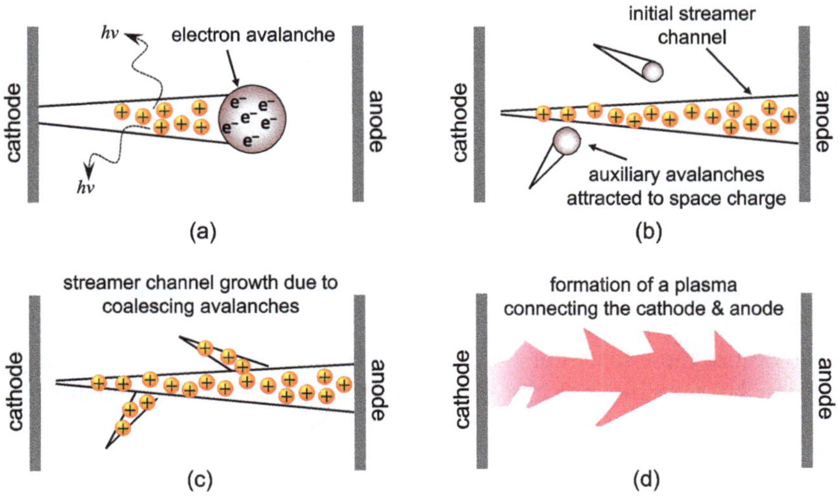

Figure 8.4. Schematic of anode directed (negative) streamer formation. (a) First an electron avalanche propagates, creating a channel of positive ions and emitting high-energy photons. (b) Additional electron avalanches are attracted to the positive space charge as the initial streamer channel reaches the anode. (c) The streamer channel grows as avalanches coalesce. (d) A conductive plasma channel is formed, connecting the anode and cathode. Adapted from [7].

and leaving positive ions in its wake. The ions, moving in the opposite direction toward the cathode, are much slower than the electrons, and thus there is some charge separation—the electrons tend to bunch together in a fast-moving group, leaving behind a slower moving collection of ions. If the ionization rate is high enough, via equation (8.10), then considerable space charge builds up and its electric field affects all subsequent avalanches. As shown in figure 8.4(b), auxiliary electron avalanches are attracted to the positive space charge of the 'quasi-stationary' positive ions. These coalescing avalanches, along with the ion motion toward the cathode, cause the ionized channel to grow, shown in figure 8.4(c), until a conductive channel fully connects the anode to the cathode. Electron emission occurs at the cathode, feeding new electrons into the ionized channel such that the process continues in a self-sustained manner, creating a plasma.

The criterion for when a streamer will form is based on the amount of charge in the ionized channel. Very roughly, if the space charge produces an electric field that exceeds the applied electric field, then ions and electrons will respond to the space charge in the electron avalanche rather than the applied electric field. Hence, subsequent electron avalanches will be attracted to the space charge region and form a streamer. Leaving the details for the reader to follow up on, the space charge field for the group of electrons at the head of the streamer can be approximated as

$$E_{SC} \approx \left(\frac{q}{R^2}\right) e^{\alpha(E_A)d}, \tag{8.18}$$

where Townsend's first ionization coefficient is evaluated at the applied electric field E_A and R is the characteristic diffusion radius for the electrons. This will exceed E_A roughly when

$$\alpha d \sim 18 - 20. \tag{8.19}$$

Equation (8.19) is known as **Meek's criterion**[12] for streamer breakdown.

The cathode directed streamer is slightly different, as the primary mechanism requires not only the electrons generating ions in the avalanche but photons generated by electron-impact excitation in the initial avalanche. These photons, as shown in figure 8.5(a), initiate avalanches by photoionization (see chapter 4). Like the anode directed streamer, these secondary avalanches are attracted to the space charge of the initial avalanche, merging with the space charge to form an ionized channel.

In both cases, streamers are highly ionized, needle-like plasma channels that move extremely fast, forming on the order of nanoseconds in a typical laboratory setting. They can also branch out to form complex structures, akin to lightning. For most experimental systems, streamers will form when $pd > 300$ Torr cm (or so). However, sharp edges, sharp tips, or any other sharp features that concentrate the electric field can cause streamers to form (sometimes unintentionally) at lower pd conditions.

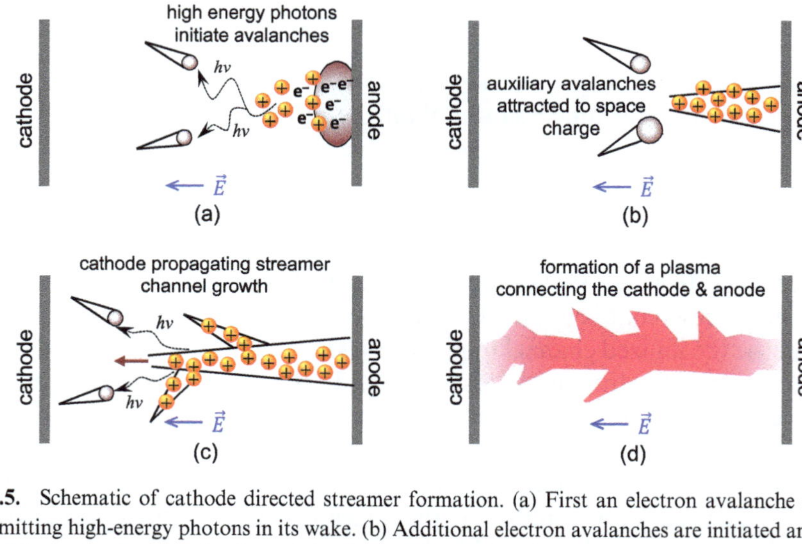

Figure 8.5. Schematic of cathode directed streamer formation. (a) First an electron avalanche reaches the anode, emitting high-energy photons in its wake. (b) Additional electron avalanches are initiated and attracted to the positive space charge in the channel that has reached the anode. (c) The streamer channel grows toward the cathode as avalanches coalesce. (d) A conductive plasma channel is formed, connecting the anode and cathode. Adapted from [7].

[12] Named after John M Meek, an American physicist who derived the criterion in 1940 [8]. Meek was one of a handful of researchers making significant advances in our understanding of streamers and gas discharges in the period following Townsend's pioneering work. Along with Meek, this included his collaborator Leonard B Loeb, another American, who co-authored the classic book *The Mechanism of Electric Spark* [9], and Heinz Raether, a German physicist who did some of the first foundational experiments on streamers and sparks [10].

8.4 Summary

The purpose of this chapter was to give an overview of the basics of discharge formation and breakdown, building upon the general principles laid out in parts I and II of this book. The two main points are that there are generally two types of breakdown, Townsend breakdown and streamer breakdown, and (with some manipulation) the breakdown criteria in equations (8.11) and (8.19) can be written as ionization criteria as:

- Townsend's criterion (Townsend breakdown): $\alpha d \sim \ln\left(1 + \frac{1}{\gamma}\right)$.
- Meek's criterion (streamer breakdown): $\alpha d \sim 18–20$.

Note that these are both for the case of a 1D geometry with a uniform electric field of $E_A = V_A/d$. However, if the field is *not uniform*, such that the electric field is a function of position r and α is a function of E, then this is accounted for by integrating Townsend's ionization coefficient over the path between the two electrodes. The criteria can thus be rewritten as

- Townsend's criterion (Townsend breakdown): $\int_0^d \alpha(E(r))dr \sim \ln\left(1 + \frac{1}{\gamma}\right)$.
- Meek's criterion (streamer breakdown): $\int_0^d \alpha(E(r))dr \sim 18 - 20$.

What has been remarkable about these criteria, and the subsequent Paschen's law in equation (8.13), is that even though discharge geometries, conditions, and configurations can take many shapes, they for the most part hold true. While they are only predictive for an idealized experimental apparatus that align with the underlying 1D approximations, the general behavior observed experimentally is consistent with the qualitative behavior predicted by these criteria. For example, one can fairly easily find examples of Paschen curves throughout the plasma and gas discharge literature, indicating that the basic non-monotonic shape with the left and right branches and local minimum is phenomenologically correct.

Finally, it is important to note that while this chapter focused on basic concepts of breakdown from the perspective of plasma generation, the study of breakdown is an incredibly rich research area on its own. For example, breakdown under ac, RF, or transient fields was not discussed nor breakdown from other stimuli such as lasers. However, even beyond its role in plasma and gas discharge devices, breakdown is an extremely relevant area in all electrical systems, as everything from MEMS to power lines will fail if they reach critical breakdown voltages. For example, as noted in section 8.2, the deviation from Paschen's law at microscale dimensions was first observed by engineers developing MEMS devices. Breakdown in electrical systems is a broad general field called *dielectric breakdown*, which not only includes gas breakdown, as discussed here, but also other dielectrics including solids such as ceramics and plastics. There are many common phenomena as the basic principle—the transition of an insulating medium to a conducting medium—is shared.

References

[1] Raizer Y P 1991 *Gas Discharge Physics* (Berlin: Springer)

[2] Paschen F 1889 Ueber die zum Funkenübergang in Luft, Wasserstoff und Kohlensäure bei verschiedenen Drucken erforderliche Potentialdifferenz *Ann. Phys.* **273** 69–96

[3] Townsend J S 1915 *Electricity in Gases* (Oxford: Clarendon)

[4] Descoeudres A, Levinsen Y, Calatroni S, Taborelli M and Wuensch W 2009 Investigation of the dc vacuum breakdown mechanism *Phys. Rev. ST Accel. Beams* **12** 092001

[5] Go D B and Venkattraman A 2014 Microscale gas breakdown: ion-enhanced field emission and the modified Paschen's curve *J. Phys. D: Appl. Phys.* **47** 503001

[6] Lisovskiy V A, Yakovin S D and Yegorenkov V D 2000 Low-pressure gas breakdown in uniform dc electric field *J. Phys. D: Appl. Phys.* **33** 2722

[7] Fridman A 2008 *Plasma Chemistry* (New York: Cambridge University Press)

[8] Meek J M 1940 A theory of spark discharge *Phys. Rev.* **57** 722–8

[9] Loeb L B and Meek J M 1941 *The Mechanism of the Electric Spark* (Stanford, CA: Stanford University Press)

[10] Raether H 1939 The development of electron avalanche in a spark channel (from observations in a cloud chamber) *Z. Phys.* **112** 464

IOP Publishing

Ionization and Ion Transport (Second Edition)

A primer for the study of gas discharges and plasmas

David B Go

Chapter 9

The basic discharge structure

Once a gas discharge makes the transition from a pre-breakdown state to a post-breakdown, self-sustained state, the discharge itself takes on a different form. Depending on the exact operation of the plasma device, the interstitial gas will consist of the plasma—the glowing, voluminous, and primarily *electroneutral* region in the electrode gap—and of sheaths—darker, thin, and space-charge-dominated regions near the electrodes and any surfaces in the device. While the glowing plasma is primarily what we picture when we think of a plasma device, the sheaths are in many ways the more critical regions, where important physical and chemical processes are occurring. Certainly, for many plasma processing technologies[1] the plasma is brought in contact with a substrate or target surface and thus the behavior at these edges of the plasma plays an important role. To that end, what exactly is a sheath and how is it different to a plasma? Why does it matter if the plasma is electroneutral (and what does that mean) and the sheath is not? Why do sheaths form? This chapter will overview the basic discharge structure consisting of a plasma and sheath and lay down the foundation for understanding plasma devices.

9.1 The sheath and the plasma

Imagine, for the time being, a collection of charged particles just existing in space, potentially along with neutral particles in the background (as in our gas discharges). The collection consists of an equal number of negative and positive charges, or in our situation, free electrons and positive ions ($N_e \approx N_i$), as shown in figure 9.1(a)[2].

[1] Plasma processing of microelectronics is one of the most well-established plasma technology areas. Plasmas are used to etch material forming transistors on a computer chip, deposit material on the computer chip, and otherwise affect surface properties. But plasma processing is used in a wide variety of other applications—both mature and emerging—ranging from sterilization of surfaces to healing of wounds, that require the plasma to be in contact with a surface.

[2] Note we are ignoring all electrodes, power supplies, circuitry, and so forth to make the plasma and are just assuming that the plasma exists.

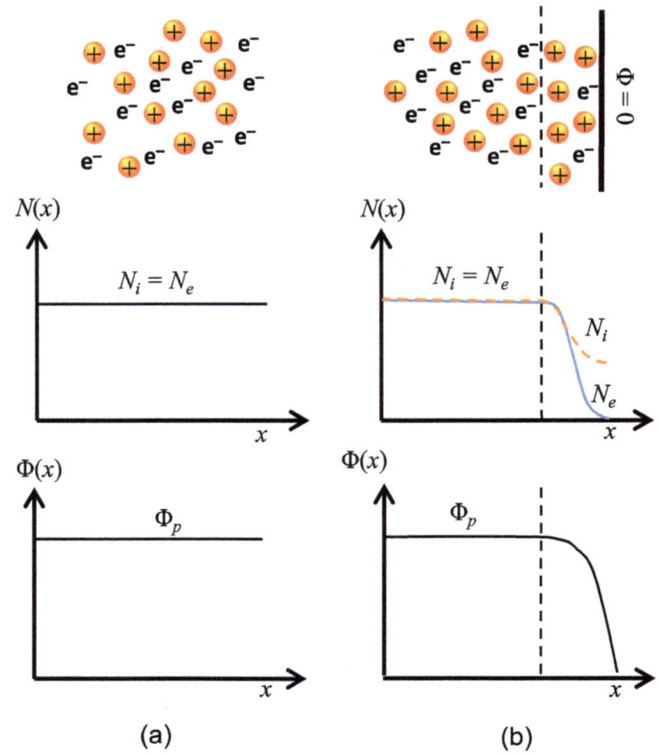

Figure 9.1. Illustration of collection of electrons and positive ions and the ensuing electron and ion charge distributions, $N_e(x)$ and $N_i(x)$, and potential distribution, $\Phi(x)$, for (a) an electroneutral plasma and (b) near a grounded plate.

Since the numbers of charges are *equal*, the collection of particles is electroneutral. Physically, this implies that there is no space charge, and Poisson's equation, equation (7.14), reduces to Laplace's equation[3],

$$\nabla^2\Phi = 0. \tag{9.1}$$

Because we are just considering a body of particles in space, there are no boundaries and thus no boundary conditions. The only solution to equation (9.1) then is that the potential field is uniform everywhere, and we can call this the plasma potential, $\Phi(x) = \Phi_p$, as shown in figure 9.1(a). As the electric field is the gradient of the potential, $\overrightarrow{E} = -\vec{\nabla}\Phi$, this also implies that there is no electric field. Thus the only motion of the electrons and ions is thermal motion[4], there is no drift component. As smaller particles and with higher energy in our non-equilibrium discharges, the electrons will diffuse faster than the ions and spread more.

[3] Named after French mathematician Pierre-Simon Laplace, and the divergence of the gradient operator ∇^2 is called the Laplacian.

[4] That is, diffusion.

Now, let us assume that we place a zero-potential, grounded plate ($\Phi = 0$)[5] at the edge of this body of particles, as shown in figure 9.1(b). The electrons will naturally move more quickly to the plate via diffusion than the slower-moving ions and be neutralized by the grounded circuit. Therefore, there will be a surplus of ions close to the plate with respect to electrons ($N_i > N_e$) whereas the remainder of the body of particles remain electroneutral ($N_i = N_e$).[6] We call this region of *positive space charge* the **sheath** while the rest of the body is the electroneutral bulk **plasma**. Now, since there is a region of positive space charge, we will form a potential field $\Phi(x)$ that is *positive* in the plasma and drops sharply to zero at the grounded plate, as shown in figure 9.1(b). This potential accelerates ions toward the plate but acts as a hill to prevent all but the most energetic electrons from moving toward the plate (they want to 'roll' back to the bulk plasma), until a balance is reached[7].

Sheaths will inherently form on any surface in contact with a plasma, whether it be an electrode, the surface of a vacuum chamber, or even particles or droplets injected into a plasma. Because of the inherent differences between fast electrons and slow plasma ions, unless the surface has an applied potential on it (as we will see in chapter 10), the surface itself will charge *negative* with electrons, and the sheath will always form a positive space charge.

9.2 The physics of the sheath

Sheaths form when there are enough electrons and ions in the plasma such that there is charge screening as charges build up in the sheath. To think about this, let us reconsider figure 9.1(b), but assume that the plate is not grounded, but unconnected to an external circuit[8]. Again, electrons will move quickly to the plate and charge it negatively such that it has a *negative* surface charge density[9]. Let us also assume that the positive ion density is so large that it is fairly uniform around the plate. The plate's negative charge *repels* additional electrons from coming to the surface creating a depletion of electrons near the plate. Again, we form a sheath of positive charge. However, at some distance the electrons in the bulk no longer feel the

[5] In this hypothetical thought experiment, we can do this. In practice, a surface is set to zero potential by 'grounding' it or connecting it to ground. What is 'ground'? Ground literally refers to the natural electric potential of the Earth, which we arbitrarily set as $\Phi = 0$ for reference. One way to do this is to drive a conductive rod into the ground (literally, dirt), where it becomes equipotential with the Earth; anything connected to the rod is thus 'grounded'.

[6] Another term that is often used is 'quasi-neutral' or $N_i \approx N_e$, which more accurately reflects the microscopic interactions and behavior of the plasma and the various processes therein.

[7] This is one of the most fascinating features of plasmas and gas discharges. This large collection of individual particles featuring all types of collisions and behaviors just naturally finds its own balance with no external forces.

[8] When we do not connect a plate to ground or to some other potential source (a battery or power supply), we typically say it is at a 'floating' potential. It will assume a potential value as charges building up on the surface until a balance with the charges in the plasma is reached and the sheath stabilizes. That is, it 'floats' to the potential it needs. If we *do* connect the plate to an external circuit, it behaves as a probe, where the current and voltage through the circuit will reflect the temperatures and charge concentrations in the plasma. The Langmuir probe is the most popular of these.

[9] Where surface charge density is often represented by σ_S and has units of C m^{-2}

repulsive nature of the plate and therefore are *screened* or *shielded* from further depletion. That is, they only feel the attractive space charge of the positive ions in the sheath.

9.2.1 Debye length

The characteristic distance over which electrons are effectively screened is called the **Debye length** (λ_{De})[10]. That is, electrons and ions in the bulk of the plasma are largely not disturbed by surfaces when they are approximately one λ_{De} away. Mathematically, the Debye length can be expressed as

$$\lambda_{\text{De}} = \left(\frac{\epsilon_0 k_b T_e}{q^2 N_e}\right)^{\frac{1}{2}}. \tag{9.2}$$

The key here is that the Debye length *decreases* as the electron density N_e increases. Let us rationalize this. As the electron concentration increases, the concomitant space charge increases, such that they more effectively screen the electric field; other free charges can come *closer* without feeling the electric field. Hence the Debye length is shorter for larger N_e.

We can get a sense of the scale of the Debye length if we assume the electrons and ions have a number density of $N_e = N_i = 10^{10}$ cm^{-3} and the electron temperature is 2 eV. This gives a Debye length of $\lambda_{\text{De}} \sim 140$ μm, or about 1/10 of a millimeter. This length actually plays an important role in even establishing the existence of a plasma. Consider again our basic system of two electrodes spaced d apart, where we apply a dc potential across them to cause breakdown and form a discharge with these properties. If our electrodes are spaced $d = 1$ cm apart, then we would expect the sheath at the electrodes to about 1/100 of our electrode gap. What happens if we make the electrode gap d even smaller and smaller, to about the size of the Debye length $\lambda_{\text{De}} \sim d$, in this case about 0.1 mm? One of two things can happen. On the one hand, our gas discharge simply will not transition into a full, quasi-neutral plasma. Electrons and ions will still be present (and thus a discharge with current across the gas will exist), but not at sufficient concentration to ever balance each other effectively and create a plasma and sheath system. Alternatively, if our power supply and circuitry allow, our system will form a plasma but with a much higher electron and ion densities, decreasing the Debye length and size of the sheath. That is, a criterion for an ionized gas discharge to be considered a plasma is that there is sufficient ionization to produce charge densities such that $\lambda_{\text{De}} \ll d$[11].

[10] It is also called the Debye radius, and to be clear, a surface does not need to be present for this screening effect. It is named after Peter Debye, a Dutch physicist and chemist who was awarded the Nobel Prize in chemistry for contributions to understanding molecular structure.

[11] Another way to think of this is that the smaller the plasma or discharge device, the higher the electron density. Over the past few decades, this has actually led to a new area of plasma science and engineering called microplasmas (or microdischarges). These are loosely described as devices with sub-millimeter electrode gaps, and very, very high electron concentrations have been observed because of the very small scale [1]. As the gaps decrease even further to the micron scale, the physics of the system will actually change [2].

9.2.2 The potential field in the sheath

One of the key features of the sheath is that there is an appreciable potential field that arises from the build-up of space charge (figure 9.1(b)), which in turn generates an appreciable electric field. This is in contrast to the electroneutral plasma, where there is no electric field. In this sense, the sheath is important because it is the portion of the discharge where both electrons and ions are accelerated, giving them increased energies affecting ensuing collisions and so forth. As we will learn in chapter 10, the sheath is what ensures that dc discharges maintain their self-sustained state. Therefore, a basic understanding of the potential difference across the sheath is important. Here we will assume a *collisionless* sheath, as it simplifies the analysis, but readers are encouraged to also learn about *collisional* sheaths, which more accurately reflect what happens in most practical gas discharges [3]. Figure 9.2 shows a more microscopic version of the sheath region, breaking it into the electroneutral bulk, a quasi-neutral pre-sheath, and then the sheath itself. As we will see, the pre-sheath is essential to the existence of a sheath itself.

To analyze the potential in the sheath using Poisson's equation, equation (7.14), we need to know the number density of electrons and ions in the sheath. To start, let us consider a body of electrons that are diffusing through some physical space. If we consider the electron momentum equation, equation (7.10b), and assume the electrons are in thermal equilibrium and *collisionless*[12], we can simplify it to

$$qN_eE_x + k_bT_e\frac{dN_e}{dx} = 0. \tag{9.3}$$

If we apply the relationship between the potential and electric field, equation (7.15), and do some rearranging, it can be shown that equation (9.3) can be rewritten as

$$\frac{d}{dx}(-q\Phi + k_bT_e\ln(N_e)) = 0. \tag{9.4}$$

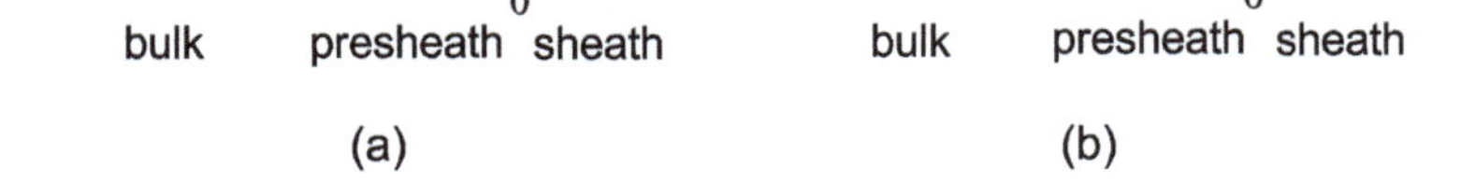

Figure 9.2. Schematics of the variation of the (a) electron and ion number densities and (b) potential field in the bulk, pre-sheath, and sheath of a plasma, assuming a collisionless sheath.

[12] Note that this is the same assumption in the drift-diffusion approximation in section 7.2, except there we assumed there are collisions.

The solution to this first-order differential equation is

$$N_e = N_{e,0}\exp\left(\frac{q\Phi(x)}{k_bT_e}\right), \tag{9.5}$$

where $N_{e,0} = N(x = 0)$. This is called the **Boltzmann equilibrium distribution** for electrons, and describes the distribution (in one-dimension) of equilibrium electrons in response to the potential field $\Phi(x)$ they build up; where there are more electrons, the magnitude of the potential is greater and vice versa. In our *collisionless* sheath, we can assume the electrons take on a Boltzmann distribution through the sheath thickness.

Ions, on the other hand, do not assume a Boltzmann distribution[13]. Instead, let us assume that ions are accelerated into the sheath by the small potential difference across the pre-sheath in figure 9.2 to have a kinetic energy of $\frac{1}{2}m_iv_i^2$. When we consider the sheath in figure 9.2, we recognize that as they enter the sheath, the electron and ion number densities are the same, implying $N_{e,0} = N_{i,0} = N_0$. Using this boundary feature, we can write Poisson's equation as

$$\frac{d^2\Phi}{dx^2} = \frac{qN_0}{\epsilon_0}\left(\exp\left(\frac{q\Phi(x)}{k_bT_e}\right) - \left(1 - \frac{2q\Phi(x)}{m_iv_i^2}\right)\right), \tag{9.6}$$

where the exponential term accounts for the Boltzmann distribution of electrons and the $\frac{2q\Phi(x)}{m_iv_i^2}$ accounts for the ions entering the sheath with an energy of $\frac{1}{2}m_iv_i^2$ and being further accelerated by the sheath potential $\Phi(x)$.

Equation (9.6) is a very non-linear, second-order differential equation and pretty challenging to solve analytically[14], but we can apply some mathematical manipulation to learn something from it. If we multiply the entire equation by $d\Phi/dx$ and integrate over x, the second-order term becomes a first-order term,

$$\left(\frac{d\Phi}{dx}\right)^2 = \frac{2qN_0}{\epsilon_0}\left(\exp\left(\frac{q\Phi(x)}{k_bT_e}\right) - \left(1 - \frac{2q\Phi(\mathrm{x})}{m_iv_i^2}\right)\right). \tag{9.7}$$

Again, this first-order differential equation is non-linear and hard to solve. But we can observe that the squared $\frac{d\Phi}{dx}$ term on the left-hand side of equation (9.7) requires that the right-hand side of the equation be positive[15] or

$$\exp\left(\frac{q\Phi(x)}{k_bT_e}\right) - \left(1 - \frac{2q\Phi(x)}{m_iv_i^2}\right) > 0. \tag{9.8}$$

[13] Though there is an equivalent Boltzmann distribution for ions at equilibrium, $N_i = N_{i,0}\exp\left(-\frac{q\Phi(x)}{k_bT_i}\right)$.

[14] Imagine the complexity if we had included collisions!

[15] Because the gradient of the potential $\frac{d\Phi}{dx}$ is the electric field, is must be a real number, hence the square is positive.

Further, if we do a Taylor expansion on the exponential term, it can be shown that we can get an expression for the ion velocity as they enter the sheath:

$$v_i \geqslant \left(\frac{k_b T_e}{m_i}\right)^{\frac{1}{2}}. \tag{9.9}$$

That is, in a collisionless sheath, ions *must* be accelerated by pre-sheath to *at least* the thermal energy of the electrons in order for a potential field to exist in the sheath. Another way to think of it, is that the plasma *must create a pre-sheath* that accelerates the ions to sustain the sheath; the pre-sheath is a necessary component for the existence of the sheath. This is again a great example of how this collection of charged particles autoregulates itself to form plasma and sheath regions.

Equation (9.9) is a limiting condition for the ions and is called the **Bohm velocity**[16],

$$v_B = \left(\frac{k_b T_e}{m_i}\right)^{\frac{1}{2}}. \tag{9.10}$$

The criterion for the (collisionless) sheath to exist, i.e. ions must be accelerated to at least the Bohm velocity, is thus called the **Bohm sheath criterion** [4, 5]. From the perspective of a non-equilibrium plasma, this is particularly interesting, as we assume the ions to be much cooler than the electrons ($T_i \ll T_e$). This may be true in the bulk of the plasma, but at the edges, near surfaces, other behavior is manifested.

9.3 Summary

The purpose of this chapter was to give an overview of the basic structure of a body of charged particles, consisting of a plasma and sheaths. The key takeaways include:

- Sheaths are regions of local space charge that form at *any* surface in contact with a plasma, including electrodes, chamber walls, and even particles or measurement probes.
- Sheaths form because of the differences in the diffusivity of electrons and ions that lead to charge separation and ultimately charge screening effects, where the distance of charge screening is called the Debye length. The size of the sheath is on the same order of magnitude as the Debye length.
- The behaviors of electrons and ions autoregulate themselves to achieve the formation of a sheath and plasma, where additional regions form (such as pre-sheaths) to ensure their existence. If the number density or energies (temperatures) of the electrons and ions are not sufficient, then a electro-neutral plasma will not form (though a discharge may still exist).

The reader should recognize that this chapter presented a very high level view of plasmas and sheaths, and the actual behavior in the sheaths of practical devices is

[16] Named after David Bohm, an English physicist.

not only more complex (i.e. involves collisions) but a very active area of ongoing research. Sheath behavior is particularly important in non-equilibrium, low-temperature devices that are used to induce some behavior on a target, such as in microelectronics fabrication, sterilization, or even biomedical applications.

It is also important to know that sheaths are not unique to plasmas, though they are often called Debye layers (as well as electric double layers) in other disciplines. They naturally arise in any system that contains mobile charge carriers and are in contact with an interface. This includes liquid systems, such as electrolytes used in batteries and electrochemical cells, as well in solid systems, such as in semiconductors and transistors[17]. The primary difference is both the charge densities and charge diffusivities in those systems are much different than those in gaseous plasmas; for example, most electrolytes consist of equally sized cations and anions as opposed to small electrons and large ions. Yet the basic behavior is still the same, making Debye layer behavior a rich area of physics unto itself.

References

[1] Schoenbach K H and Becker K 2016 20 years of microplasma research: a status report *Euro. Phys. J.* D **70** 1–22

[2] Bilici M A, Haase J R, Boyle C R, Go D B and Sankaran R M 2016 The smooth transition from field emission to a self-sustained plasma in microscale electrode gaps at atmospheric pressure *J. Appl. Phys.* **119** 223301

[3] Oksuz L and Hershkowitz N 2005 Plasma, presheath, collisional sheath and collisionless sheath potential profiles in weakly ionized, weakly collisional plasma *Plasma Sources Sci. Technol.* **14** 201–8

[4] Bohm D 1949 *The Characteristics of Electrical Discharges in Magnetic Fields* ed A Guthry and R K Wakerling (New York: McGraw-Hill)

[5] Riemann K-U 1991 The Bohm criterion and sheath formation *J. Phys. D: Appl. Phys.* **24** 493–518

[6] Rumbach P, Clarke J P and Go D B 2017 Electrostatic Debye layer formed at a plasma–liquid interface *Phys. Rev.* E **95** 053203

[7] Elg D T, Delgado H E, Martin D C, Sankaran R M, Rumbach P, Bartels D M and Go D B 2021 Recent advances in understanding the role of solvated electrons at the plasma–liquid interface of solution-based gas discharges *Spectrochim. Acta* B **186** 106307

[8] Tchertchian P A, Wagner C J, Houlahan T J Jr, Li B, Sievers D J and Eden J G 2011 Control of the interface between electron–hole and electron–ion plasmas: hybrid semiconductor-gas phase devices as a gateway for plasma science *Contrib. Plasma Phys.* **51** 889–905

[17] An exciting new direction for plasmas are when they interface with these other systems, such as in plasmas in contact with liquids or plasmas in contact with semiconductors. The reader is encouraged to learn more about these areas starting with [6–8].

IOP Publishing

Ionization and Ion Transport (Second Edition)
A primer for the study of gas discharges and plasmas
David B Go

Chapter 10

The classical dc discharge

The transition of the gas from an insulator to a conductor is achieved once the breakdown voltage (or electric field) is reached. Then what happens? That largely depends on the specifics of the actual plasma device and how it is operated. First, as was observed in chapter 8, the equations for Townsend discharge predict that the current density will increase to infinity as the breakdown voltage is approached. This is clearly not physically possible and, in practice, the current will be limited by the power supply and the circuit that form the plasma device. In most cases, if nothing is used to intentionally limit the current, as the voltage surpasses the breakdown threshold, the rapid growth in current will form a highly transient spark. This spark will be extinguished by reaching the limits of what the power supply can provide or by other physical mechanisms, such as heating of the gas causing it to rise (like in a Jacob's ladder configuration) or, as we will learn, because the voltage across the gap drops. What if we do not want the discharge to form a spark and instead form a stable, non-equilibrium, low-temperature discharge? And once it does, what is the nature of that discharge? This chapter will overview the most basic approach to forming such a stable plasma using a dc circuit and overview its basic properties.

10.1 The basics of electrical circuits

Before we describe the formation of a glow discharge, it is a good idea to review the basics of electrical circuits, because these will be important for understanding a plasma circuit, which is just an electrical circuit with a plasma included. An electrical circuit is typically constructed of an electrical source, such as a power supply that provides a potential difference and current across the circuit, and various elements that regulate either the current or voltage. The three most common of these are the resistor, capacitor, and inductor, which are shown in figure 10.1.

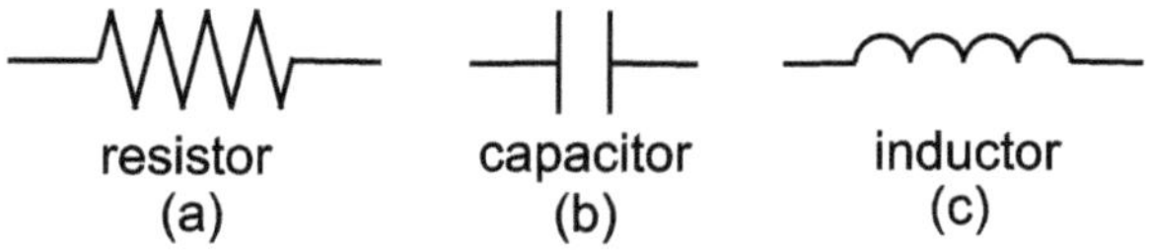

Figure 10.1. Standard symbols for (a) resistor, (b) capacitor, and (c) inductor circuit elements.

A resistor limits current through it ('resisting'), but still allows charge to pass. The relationship between current (i) and potential (Φ) most often follows **Ohm's law**[1]

$$\Phi = iR, \tag{10.1}$$

where R is the resistance of the circuit element in units of ohms. Note that if the potential across the resistor (also called the voltage drop) is fixed, when R is small, the current will be large, and when R is large, the current will be small. Metal wires that make up basic electrical devices are conductors, which implies that they have very low resistance. When a discharge becomes conductive, it has *low resistance*, and can pass a fair amount of current[2].

A capacitor does not allow current to pass through it; instead it builds up the current coming from the power supply and 'stores' that charge. If the power supply is dc, the charge will build up on the capacitor until the current either extinguishes or the capacitor breaks down[3] and starts flowing current. However, in an ac circuit, the charge will build up when the potential difference is in one direction and then flow the opposite way once the potential difference switches direction. This transient behavior is captured by a time-dependent current and potential relationship,

$$C\frac{d\Phi}{dt} = i, \tag{10.2}$$

where C is the capacitance and measured in farads[4]. For example, if the potential has the form of $\Phi(t) = \Phi_0 \sin(\omega t)$, the current will have the form of $i(t) = i_0 \cos(\omega t)$, where i_0 is related to the capacitance C and frequency ω.

Physically, the capacitance is the amount of charge that can be held across the capacitor for a given potential difference before it breaks down. A capacitor is typically made by two electrodes of surface area A sandwiching an insulating material of thickness d and permittivity ϵ, such that the capacitance can be defined as

$$C = \frac{\epsilon A}{d}. \tag{10.3}$$

---

[1] Named after German physicist Georg Simon Ohm, and whose units bear his name 'ohm' and the symbol Ω, where $1\ \Omega = 1\ \mathrm{V\ A^{-1}}$.

[2] However, as shown in figure 10.3, the current–voltage behavior for a dc discharge is not linear and decidedly not Ohmic.

[3] Just like gas breakdown, as discussed in section 8.4.

[4] Named after the esteemed English physicist Michael Faraday and taking the symbol F, where $1\ \mathrm{F} = 1\ \mathrm{C\ V^{-1}}$.

As gases are naturally insulating, they are often used as capacitors. In fact, the basic plasma device described in chapter 8, consisting of two electrodes separated by a gas gap, is essentially a capacitor that transitions to a resistor once breakdown is achieved.

An inductor is like a capacitor in that it stores electrical energy, but rather than in the form of charge, it is stored in a magnetic field. It is formed by an insulated wire wound into a coil shape; when current passing through the wire varies in time, it induces a magnetic field[5] and a voltage that opposes the current. Like a resistor, an inductor will limit current, but only if it is time varying as with an ac power supply, such that the relationship between current and potential is

$$\Phi = L\frac{di}{dt}, \tag{10.4}$$

where L is the inductance and measured in henrys[6]. While resistors and capacitors are often used as equivalent circuit elements for a plasma, inductors are typically not, though they can be important in the circuit used to form and drive ac plasma devices (including inductively coupled plasmas).

10.2 dc discharge circuit and operation

Figure 10.2(a) shows a typical dc discharge circuit. The power supply provides an applied voltage V_A and current i first passes through a resistor with resistance R and then into our discharge device, which is simply two electrodes separated by a gap d in a chamber maintained at pressure p. If no discharge is generated (and there is no

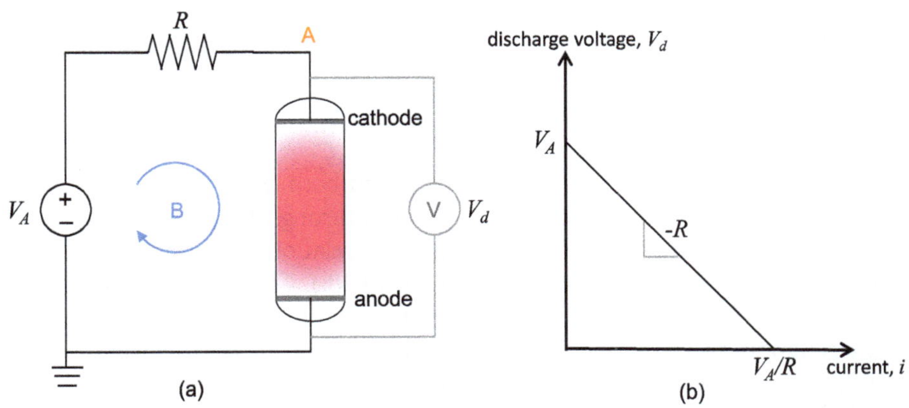

Figure 10.2. (a) Schematic of circuit for conventional dc discharge system, including a power supply providing applied voltage V_A, resistor with resistance R, and discharge device consisting of two electrodes separated by a distance d and in a chamber maintained at a pressure p. A voltmeter attached across the electrodes can measure the discharge voltage V_d. (b) Illustration of load line for the discharge device relating the discharge voltage V_d and discharge current i. At zero current ($i = 0$), the voltage across the discharge device is equivalent to the applied voltage V_A.

[5] This is known as Faraday's law of induction.

[6] Named after the American Joseph Henry and taking the symbol H, where 1 H = 1 Ω s.

plasma), the discharge device is a capacitor, whose capacitance can be determined by equation (10.3). To analyze this circuit, we can use two laws called Kirchhoff's current law (also called his first law) and Kirchhoff's voltage law (or second law)[7]. Kirchhoff's current law states that current (or more accurately charge) must be conserved in any node in a circuit. That is, we can take any node, such as the one labeled A in figure 10.2, and determine that the amount of current going into the node must equal the amount of current going out or

$$\sum_{\text{node}} i_{in} = \sum_{\text{node}} i_{\text{out}}. \tag{10.5}$$

As noted above, for a given current (or steady dc or transient ac), there will be some potential difference or voltage drop across each circuit element. Kirchhoff's voltage law states that energy must be conserved in a closed circuit loop. That is, the sum of all potential differences around an entire loop[8] in the circuit, such as the one labeled B in figure 10.2, must be zero or

$$\sum_{\text{loop}} \Phi_{\text{element}} = 0. \tag{10.6}$$

Both the resistor and the plasma device in our circuit will drop some voltage across them, V_R and V_d (for discharge), respectively. While the applied voltage is known and the voltage drop across the resistor will be considered shortly, the voltage across the discharge can be measured with a voltmeter with leads attached to the anode and cathode respectively. Applying Kirchhoff's voltage law around the circuit's loop gives

$$V_A - V_R - V_d = 0. \tag{10.7}$$

If we apply Ohm's law from equation (10.1) to the resistor and do some simple rearranging, we can rewrite this as

$$V_d = V_A - iR. \tag{10.8}$$

That is, the voltage across the discharge depends on both the applied voltage and the resistance of the resistor. As the current that flows through the resistor will also flow through the discharge in this circuit, we can plot the relationship between the discharge voltage and discharge current as shown in figure 10.2(b). This linear relationship is the load line for the circuit with a slope of $-R$. At the limit of no plasma device, the x-intercept is simply the current $i = V_A/R$ and at the limit of no (or extremely small) current, the voltage across the discharge is the applied voltage $V_d = V_A$.

Another way to view this circuit is by rearranging equation (10.8) to solve for the discharge current

$$i = \frac{V_A - V_d}{R}. \tag{10.9}$$

[7] Named after German physicist Gustav Kirchhoff.

[8] A loop is a closed path that starts at one node and returns to the original node without passing any other nodes more than once. Figure 10.2 shows only a single loop, but more complex circuits can have more than one loop. The Wheatstone bridge is perhaps the most common example.

The discharge current not only depends on the applied voltage and voltage across the plasma device, but also the resistor. For this reason, we call the resistor in this circuit a **ballast resistor** because it limits the current in the discharge. If the resistor has a large resistance, the current through the discharge is small and *vice versa.* Electrical ballasts, more generally, can be various combinations of resistors, capacitors, and inductors and are important in plasma circuits because they limit the current that flows once breakdown is achieved. This prevents the discharge from overheating and transitioning into a high current density spark.

10.2.1 The canonical current–voltage curve

Phenomenologically, dc discharges follow a canonical current–voltage curve with qualitative behaviors that can be attributed to different states of the discharge, shown in figure 10.3. It is often written with the current on the x-axis and the voltage on the y-axis, even though intuitively, the power supply controls the voltage and not the current (that is the voltage is the independent variable). This will become clear momentarily.

First consider a basic plasma circuit, where the power supply is slowly increased from $V_A = 0$ V (the y-axis). Initially, there is very little current, consisting only of electrons created by background radiation and the few ionization events they produce. But as discussed in section 8.1, the current grows as the voltage is increased, because Townsend's first ionization coefficient α is a function of the applied voltage. As the current density expressions are exponential with α, the current will grow rapidly once the voltage begins to approach the breakdown voltage, V_b. We call this behavior a **dark discharge** or **Townsend discharge** as it is largely dictated by normal Townsend processes, yet the number density of electrons it is not at sufficient that there is a visible plasma. Once the breakdown voltage V_b is achieved, there is a rapid increase in both

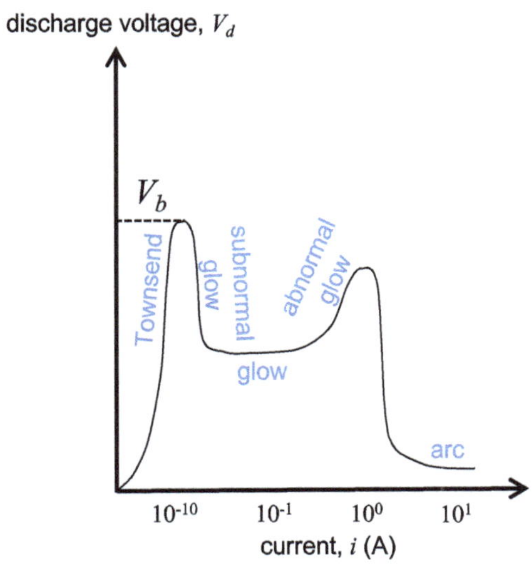

Figure 10.3. Canonical current–voltage curve for a dc discharge illustrating the different discharge regimes for different currents and the corresponding discharge voltage.

current and charge density (on the order of magnitudes, as denoted by the log scale) such that the ballast resistor now takes effect and limits the current flowing through the circuit. At this point, there is sufficient charge density in the electrode gap that charge screening becomes important and a plasma with sheaths forms (see chapter 9). What is interesting is that as the current is increased further, the voltage across the discharge actually drops very steeply, forming what is often called a **subnormal glow discharge**. As current is increased further, we then begin to form what is known as a **glow discharge** (or normal glow discharge), which is exactly what it sounds like—a stable, glowing plasma. This is the most classic discharge mode, and as will be discussed in the following sections, has some very interesting features. What is most noticeable on the canonical current–voltage plot is that it flattens. That is, the current can be increased, but the discharge voltage stays the same. Notably, this can occur over a few orders of magnitude of current. Once the current is too high, the voltage begins to climb again and an **abnormal glow discharge** is formed. Beyond the abnormal glow discharge, the rate of ionization increases appreciably, with a commensurate increase in current, such that the plasma begins to *thermalize* and transition into an equilibrium plasma called an arc. The voltage drop across an arc is very small, but the current (and power) are very high, making the plasma very hot, and thus useful in applications such as arc discharge machining.

10.2.2 The discharge circuit

So how does this curve relate to the dc circuit in figure 10.2 or, how does the circuit manifest itself in such a curve? We can *overlay* the load line in figure 10.2(b) and where this load line intersects the phenomenological curve is the point of operation for the device, as shown in figure 10.4.

First consider the case where the applied voltage V_A is increased, but below the breakdown voltage, $V_A < V_b$. As shown in figure 10.4(a), the load line (with slope of $-R$) intersects the canonical current–voltage curve at a point to the left of the

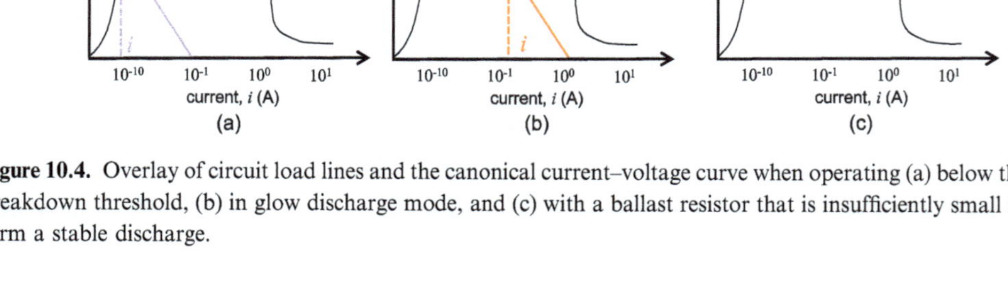

Figure 10.4. Overlay of circuit load lines and the canonical current–voltage curve when operating (a) below the breakdown threshold, (b) in glow discharge mode, and (c) with a ballast resistor that is insufficiently small to form a stable discharge.

breakdown point. If we draw a dotted line from the intersection point to the discharge voltage axis (y-axis) and current axis (x-axis), we observe that the discharge voltage is less than the applied voltage and the current is very low (typically less than nanoamps[9]). Practically, a typical ballast resistor could be ~100 kΩ. If the current is in nanoamps, then the potential difference across the ballast resistor, $V_R = iR$, is on the order of microvolts. Thus for all intents and purposes, from equation (10.8) we see that $V_d \approx V_A$, though the figure does not resolve how close these two values may be.

As the voltage is increased more, eventually it reaches the breakdown voltage, $V_A = V_b$, which is typically on the order of several hundred volts or kilovolts. This transition is very fast and in a practical experiment the current jumps dramatically—orders of magnitude—as the discharge transitions into the subnormal or normal glow regimes. Under many circumstances, the subnormal glow regime is challenging to stabilize (and can even have its own oscillations [1]), and the discharge will immediately transition into a glow. This corresponds to figure 10.4(b). There are several nuances to observe. Now the corresponding current is much higher than it was below breakdown by several orders of magnitude, and practical values can range from milliamps to amps depending the exact configuration. Correspondingly, the potential difference across the ballast resistor can be tens or even hundreds of volts. Per equation (10.8), the discharge voltage is now much lower than the applied voltage from the power supply, corresponding to the intersection on the y-axis.

Another interesting feature of the glow discharge regime is that the canonical curve is flat. Since the slope of the load line is fixed by the ballast resistor, if the applied voltage is increased further. While the current increases with each increment, the discharge voltage remains roughly the same. Thus the power supply, which was acting as a voltage source by changing the potential across the electrodes when we were *below breakdown*, is now acting as a current source by changing the current across the electrodes *after breakdown*. This behavior is entirely enabled by the ballast resistor, and the magnitude of the ballast resistor will affect exactly how the power supply behaves as a current source.

To illustrate the impact of the ballast resitance, notice that as R decreases, the slope decreases, and the load line becomes more and more horizontal as shown in figure 10.4(c). If the ballast resistor is not large enough, then a stable glow discharge cannot be formed; the load line essentially does not *intersect* the canonical curve. In this case, as the applied voltage is increased beyond the breakdown voltage, what is known as overvoltage, the discharge will transition directly to an arc, or more realistically, a highly transient spark, until it hits the limits of the power supply and extinguishes.

It is important for us to understand that the canonical current–voltage curve is phenomenological in nature; it has been observed under laboratory conditions and effectively captures the phenomena, but is not easily derived from first principles. Therefore, the exact behavior depends on the specific parameters of an experiment, including the shape and material of the electrodes, electrode gap, pressure, gas (or

[9] Recall that 1 nA = 10^{-9} A. As a point of calibration, most outlets in a typical home provide up to 15 A of ac current and a typical toaster, for example, draws roughly 5–10 A. So less than 1 nA is an *extremely* small amount of current and often requires a specialized instrument called a picoammeter to measure it.

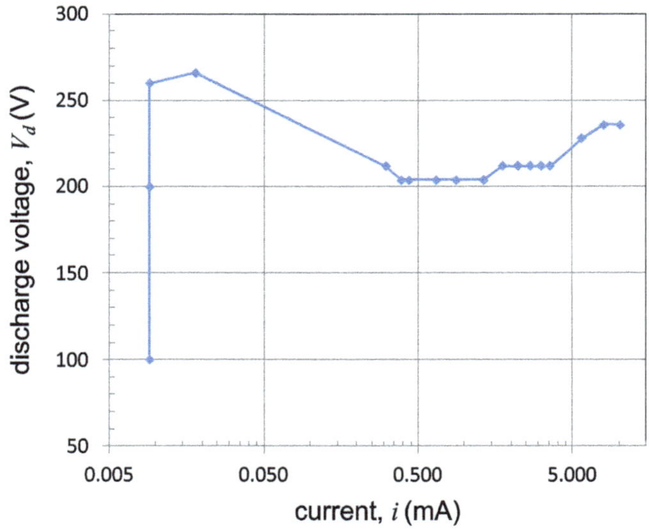

Figure 10.5. Experimental current–voltage curve for a glow discharge conducted with two copper electrodes spaced $d = 1$ cm apart and in $p = 0.9$ Torr air using a ballast resistor of 220 kΩ.

gas mixture), and so forth. The current–voltage curve is not known *a priori*, and is more or less discovered experimentally.

Figure 10.5 shows a characteristic, experimentally obtained current–voltage response between two copper electrodes at a distance of 1 cm in air at 0.9 Torr, which gives a *pd* of 0.9 Torr cm, just at the lower edge of typical operating range. Note that below breakdown, the current is very low (less than 0.01 mA), but when the voltage reaches its breakdown value of roughly 270 V, the next data point shows a nearly two order of magnitude increase in current and a discharge voltage that decreased to 212 V. As the applied voltage is increased from 280 V to nearly 2500 V, the current increases almost another two orders of magnitude, but the measured discharge voltage stays roughly the same, ranging from 204 to 236 V. We can predict the breakdown voltage using Paschen's law, equation (8.13), except that we do not know the value of γ for these electrodes, which not only depends on their copper material but surface properties as well. However, a value of $\gamma = 0.02$ produces a predicted breakdown voltage of 266 V, consistent with the experimental data[10].

10.3 Properties of the dc glow discharge

Now that we understand the transition between the pre- and post-breakdown states and the existence of sheaths, we can now use these to better understand the glow discharge (and specifically the relative flat current–voltage behavior in figure 10.3).

[10] To calculate the breakdown voltage using equation (8.13), the constants A and B for air are needed. For this experiment, $pd = 0.9$ Torr cm and close to the breakdown voltage, the reduced electric field is about $E/p = 275$ V Torr^{-1} cm^{-1}. Over the range of $E/p = 100$–800 V Torr^{-1} cm^{-1}, the constants are $A = 15$ Torr^{-1} cm^{-1} and $B = 365$ V Torr^{-1} cm^{-1}, respectively [2].

Figure 10.6 shows the basic structure of the glow discharge. Note that it consists of several different regions, some which luminesce greatly and some which give very little luminescence. That is, although the glow discharge looks uniform to the naked eye, it is actually not very uniform at all.

If we begin at the cathode, we start with a sheath that is known as the **cathode fall**. While we presented the sheath in section 9.3 as a simple build-up of space charge, it is actually much more complex. In the case of a glow discharge, the cathode is an active surface with an applied potential that affects the potential distribution, electric field, and electron and ion energies and densities in the sheath. These in turn affect the electron dynamics, thus leading to three sub-regions: the Aston dark space, the cathode glow, and the cathode (or Crooke's) dark space. These regions are named because of the intensity of the luminescence from them. After the cathode fall comes the bulk of the plasma. First is the **negative glow**, which is the most luminescent (brightest) portion of the discharge. Electrons created at the cathode and in the cathode fall via ionization are accelerated due to the large electric field

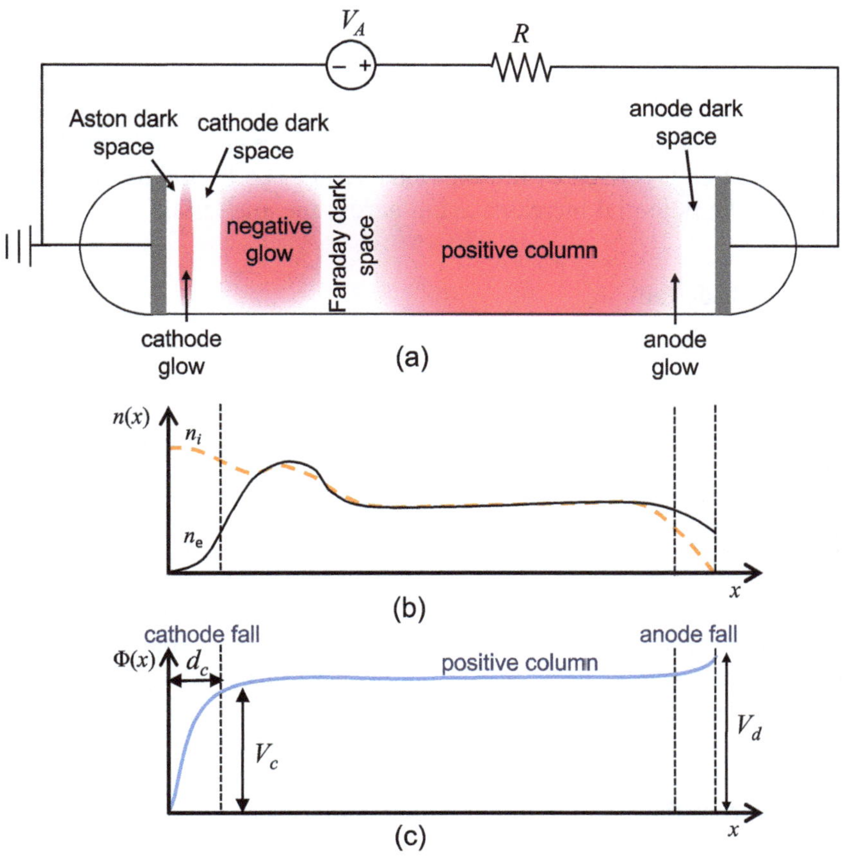

Figure 10.6. Schematic illustration of the structure of a dc glow discharge with (a) luminescent and dark regions in the discharge, (b) the electric potential distribution across the discharge, and (c) the distribution of the electron, n_e, and ion, n_i, charge densities across the discharge.

and as they exit the cathode fall they undergo a significant number of collisions creating ions and excited states, resulting in the bright glow. The bright negative glow is then followed by a region called the Faraday dark space where the electrons (having lost their energy via collisions in the negative glow) are reaccelerated, but at insufficient energy to generate a large population of excited states. The Faraday dark space is followed by the **positive column**, another luminous section of the glow that is less intense but larger in volume than the negative glow regime. At the anode, there is another sheath called the **anode fall**.

The corresponding electron and ion densities and potential field across the discharge are sketched in figures 10.6(b) and 10.6(c), respectively. As expected, there is build-up of positive space charge in the cathode fall and a corresponding steep rise in the potential difference, as is typical of most sheaths. In the large plasma regions of the negative glow and positive column, the ion and electron distributions are approximately the same, i.e. electroneutral, and the corresponding potential distribution is relatively flat, similar to what was illustrated in figure 9.1(b). That being said, there is a small voltage rise, particularly across the positive column, due to the applied voltage that is needed to sustain the glow. Unlike typical sheaths that form because of electron diffusion, in the anode fall electrons are driven to the anode by the overall electric field, while ions are driven to the positive column and ultimately the cathode. Thus the anode fall has a net negative space charge of electrons, whereas most sheaths have a net positive space charge of ions. Because the potential at the anode is fixed by the external circuit to be V_d and due to the negative space charge, the potential increases again across the anode fall.

10.3.1 The cathode fall

In many respects, the cathode fall is the most interesting part of the glow discharge. As shown in figure 10.6(c), the potential across the cathode fall V_c is nearly the same as the entire discharge voltage V_d between the cathode and anode. For this reason, we often say the cathode fall 'drops' the majority of the voltage in the discharge. Notably, the width of the cathode fall d_c is much smaller than the entire discharge length d between the anode and cathode. It is clear than that the electric field across the cathode fall is going to be very large. Furthermore, the cathode fall is bordered by a source of electrons via secondary emission at the cathode. This source of electrons and the strong electric field they experience is why the cathode fall is where much of the action happens in the discharge. Electrons emitted from the cathode will be accelerated by the electric field, both exciting and ionizing the gas in the cathode fall. Many of these electrons will maintain their high energy as they continue into the negative glow and positive column. This actually contributes to the two-temperature behavior of the electron energy distribution discussed in section 3.3; one population of fast electrons from the cathode fall and then lower energy electrons in the positive column [3]. In fact, it is the cathode fall that essentially sustains the dc glow discharge, creating enough ionization to meet the Townsend breakdown criterion in chapter 8. It is also why there is the peculiar behavior in the canonical current–voltage curve in figure 10.3 where the voltage remains the same as the current increases.

If we analyze the potential change V_c across the cathode fall [2], it can be shown to be a function of the Townsend ionization coefficients A and B, the secondary emission coefficient γ, and importantly, the product of the pressure and the distance across the cathode fall pd_c,

$$V_c = f(pd_c, \gamma, A, B).$$

This is directly analogous to Paschen's law for breakdown in equation (8.13) where the breakdown voltage is also a function of the Townsend ionization coefficients A and B, secondary emission coefficient γ, and the product pd. That is, the voltage across the cathode fall is equivalently the voltage needed to breakdown the gas, and the cathode fall distance d_c adjusts itself to ensure that the entire discharge is self-sustained. This typically corresponds to the minimum point[11] in the Paschen curve, which implies that the cathode fall self-adjusts to *minimize* the energy (i.e. potential difference V_c) required to sustain the glow discharge. This is the basic principle behind the flat portion of the canonical current–voltage curve—the discharge voltage in the glow regime V_d reflects the minimum energy needed to sustain the plasma, so it does not vary.

The natural question, then, is how does the current increase while the voltage remains the same? The nuance is in the difference between the current and current density. The minimum in the V_c versus pd_c relationship also corresponds to a minimum in the relationship between the current density j through the cathode fall and V_c. In order to maintain the energy minimization principle, the current density must also remain constant. However, recall that the current density is related to the current by the area of the discharge, or $j = i/A$. Hence, the current can grow and the current density remain constant as long as the discharge area also grows. When the glow first initiates, it typically only covers a portion of the electrodes. As the current is increased, the glow spreads radially to cover more of the surface of the electrodes, while maintaining a constant j and constant V_c (which is roughly V_d). We can visualize this graphically in figure 10.7. Figure 10.7(a) shows the relationship

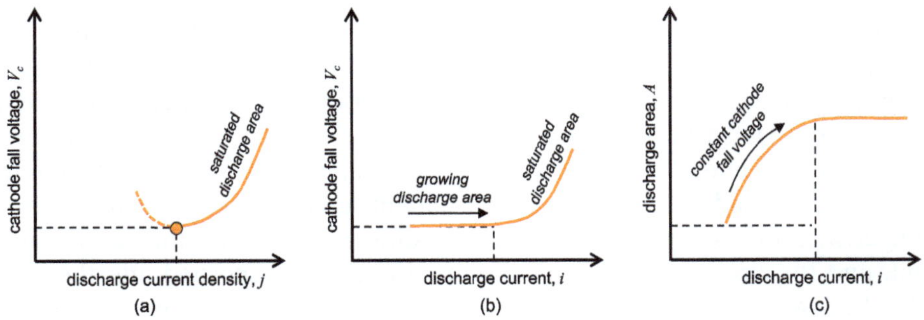

Figure 10.7. Illustrations of the relationships between cathode fall voltage, discharge current density, and discharge current, and discharge area for a glow discharge. The dot in (a) reflects the minimum point that corresponds to the operating point in the glow discharge.

[11] Also called Stoletov's point in chapter 8.

between V_c and j, with the characteristic minimum identified by the dot. When the current increases as shown in figure 10.7(b) and the area grows as shown in figure 10.7(c), the current density remains at the minimum point in figure 10.7(a) such that the cathode fall voltage is constant. Once the physical area of the cathode saturates and the discharge can no longer spread, the cathode fall voltage and current density both increase as the current increases, reflecting a transition into the abnormal glow regime. Hence, the flat portion of the canonical current–voltage curve reflects the operation of a glow discharge that only partially covers the electrodes and increasingly spreads until achieving full coverage.

10.4 Summary

The purpose of this chapter was to give an overview of the basics of dc discharge formation, from both the practical consideration of the circuit through the relationship between the current and voltage. A basic electrode configuration consisting of two electrodes and a dc power supply are considered. The key takeaways include:

- There is an experimentally observed basic behavior for the relationship between current and voltage for these configurations. The discharge progresses through several regimes spanning many orders of magnitude of current in a highly non-linear manner.
- The exact behavior of the dc discharge depends on the external circuit including the electrical ballast circuit elements that prevent spark formation.
- Once a dc glow discharge is formed, it exhibits two sheaths, a cathode fall and an anode fall, as well as a wide variety of substructures. The behavior of the cathode sheath is responsible for the flat current–voltage behavior of a glow discharge.

It is important to understand that this chapter presented the simplest of discharge configurations—a dc device with two electrodes at a given pressure and distance—that results in the canonical curve and the glow discharge. While the glow discharge remains a popular plasma operating mode from a fundamental perspective, many other modes are much more common in practical plasma devices, including high frequency plasmas that use either sinusoidal or pulsed power sources at frequencies ranging from kHz to MHz. However, the basic principles in this chapter are still useful in terms of understanding gas discharge and plasma behavior. All laboratory plasmas are beholden to the electrical circuit used to form them, from the characteristics of the power supply to the different elements that make up the circuit, and the circuit ultimately determines the energy deposited into the plasma itself. Thus, while forming a gas discharge is conceptually relatively simple, only requiring a power supply with sufficient voltage capabilities, achieving a specific plasma state or specific plasma properties (such as a desired electron energy distribution), requires much more nuance in the circuit design.

References

[1] Stefanović I, Kuschel T, Škoro N, Marić D, Petrović Z L and Winter J 2011 Oscillation modes of direct current microdischarges with parallel-plate geometry *J. Appl. Phys.* **110** 083310

[2] Raizer Y P 1991 *Gas Discharge Physics* (Berlin: Springer)

[3] Surendra M, Graves D B and Jellum G M 1990 Self-consistent model of a direct-current glow discharge: treatment of fast electrons *Phys. Rev.* A **41** 1112–24

Chapter 11

Whence and whither

As this text concludes, it is perhaps instructive to reflect on what was learned, its context in the study of non-equilibrium, low-temperature gas discharges, and what has yet to be learned to become conversant on the topic.

11.1 What was learned

The purpose of this text was really two-fold. First, it was to introduce the reader to the fundamental physics that are at the heart of gas discharges that they may not have come across in any of their prior studies. As we found in part I, these included the basics of statistical mechanics and gas dynamics in chapter 2 that were essential to understanding the behavior of free electrons in a gas (chapter 3) and the production of ions in a gas (chapter 4), and the basics of solid state physics in understanding electron emission from electrodes (chapter 5). Similarly, in part II we needed to cover the basics of transport at both the continuum and kinetic scales in chapter 6 in order to describe discharges with conservation equations (chapter 7).

The second aim was to provide the reader with a general understanding of the basic, important underlying processes that occur in and the basic fundamental properties of a gas discharge. For example, parts I and II showed that discharges are produced because high energy electrons collide with neutral particles to create ions—the so-called electron-impact ionization or α-process. Additionally, in order to replenish the gas with electrons as some leave the discharge through the anode, electrons are injected electron emission from the cathode—the so-called secondary emission or γ-process. These electrons and the ions they produce will be accelerated by the electric field produced by applying a voltage between two electrodes, causing them to drift. However, as they disperse throughout the interstitial gas they will also diffuse to oppose local gradients that build up between the electrodes. Furthermore, as the electrons and ions collide with neutral gas particles, they exchange momentum and energy, and in the case of electrons, can excite and ionize the neutral gas particles via inelastic collisions.

doi:10.1088/978-0-7503-3991-9ch11

Part III brought these ideas together to show how the α-process and γ-process are critically matched at the breakdown voltage, which leads to a self-sustained discharge (chapter 8). Once self-sustained, an electroneutral plasma is formed where in the bulk of the plasma there are an equal number of electrons and ions, but much different behavior is encountered at the edges of the plasma where sheaths are formed (chapter 9). When considering the simple dc discharge configuration, this transition to self-sustained forms a glow discharge, where the behavior in the cathode sheath, called the cathode fall, manifests in important current and voltage behavior (chapter 10).

11.2 What is next

While this primer introduced some of the key ideas that are essential to understanding non-equilibrium, low-temperature gas discharges and plasmas, it should clear that these basic concepts are not sufficient for a deep understanding of this fascinating topic. Many other important processes were neglected in the discussion presented here. For example, gas discharges and plasmas produce remarkable chemistry that relies on neutral–neutral and excited–neutral reactions that were not covered. Similarly, while the basic structure of plasmas, ranging from their quasi-neutral bulk to the sheaths they form at boundaries, was discussed, they were only touched upon at a very high level and many other characteristics (such as the plasma frequency) and processes (such as ambipolar diffusion) were not covered. Finally, the processes discussed here were presented from the perspective of a simple dc discharge, where a dc voltage is applied to two electrodes. Clearly, a more complete picture would include discussion of specific types of discharges and plasmas and the unique behaviors to each of these. Therefore, it is strongly recommended that the reader not stop here. Instead, they should use this text as a leaping-off point and proceed to learn from the numerous available texts on plasmas and gas discharges, so that they may become, in time, knowledgeable and well-versed in the field.

www.ingramcontent.com/pod-product-compliance
Ingram Content Group UK Ltd.
Pitfield, Milton Keynes, MK11 3LW, UK
UKHW051850100926
8556IPUK00005B/554